Lecture Notes in Mathematics

Edited by A. Dold and B. Eckmann

965

Topics in Numerical Analysis

Proceedings of the S.E.R.C. Summer School,
Lancaster, July 19 – August 21, 1981

Edited by P. R. Turner

Springer-Verlag
Berlin Heidelberg New York 1982

Editor

Peter R. Turner
Department of Mathematics, University of Lancaster
Bailrigg, Lancaster, LA1 4YL, England

AMS Subject Classifications (1980): 65D05, 65D07, 65H05, 65H10, 65R20, 65N30

ISBN 3-540-11967-1 Springer-Verlag Berlin Heidelberg New York
ISBN 0-387-11967-1 Springer-Verlag New York Heidelberg Berlin

Printing and binding: Beltz Offsetdruck, Hemsbach/Bergstr.
2146/3140-543210

Math

s-ep

PREFACE

THE S.E.R.C. NUMERICAL ANALYSIS SUMMER SCHOOL AND WORKSHOP

University of Lancaster

19th July - 21st August, 1981.

This meeting was arranged as a result of an initiative of the Numerical Analysis Panel of the SERC with the principal purpose of gathering together young numerical analysts and a team of experts in different areas of the subject for a period of intensive study and research. Much of the time was available for participants to pursue their own research in a stimulating environment where the experience and background knowledge of the internationally renowned experts could be drawn upon.

The other major activity of the summer school was the programme of lecture courses and seminars which occupied the beginning and end of four days each week. This programme consisted of four principal courses of ten lectures each, five shorter courses of four or five lectures each and a series of research seminars given by invited experts and offered by participants. Details of the programme are listed separately.

The aim of the principal courses was to enable someone with initially only a nodding acquaintance with a particular topic to study intensively and end up feeling competent in it while someone with a good initial knowledge would be enabled to work on significant unsolved problems in the area. These courses and the shorter courses resulted in a programme with sufficient breadth and depth that it catered for almost any specialist interest in numerical analysis and included much material to enhance participants' knowledge and understanding in other areas.

Lecture notes from most of these courses are included in this volume. Two of the exceptions, the contributions of Parlett and Gill are based on recently completed books,

B.N. Parlett, "The Symmetric Eigenvalue Problem", Prentice Hall, 1980

and

P.E. Gill, W. Murray and M. Wright, Practical Optimization", Academic Press, 1981.

This volume is thus an account of the proceedings of a very different sort of conference from those with which we are more familiar, but one which will probably become increasingly necessary as individual knowledge becomes more and more specialized.

Acknowledgements

My first and most important acknowledgement is to the Science and Engineering Research Council who sponsored the meeting to the extent of all the organizational and running costs and a 50% contribution to the expenses of U.K. participants.

Without this financial backing the event would not have taken place. I also wish to thank the Mathematics Secretariate of the SERC, the members of the Numerical Analysis Panel in general and the organizing committee (Professors C.W. Clenshaw, K.W. Morton and M.J.D. Powell) in particular for all their assistance and encouragement in making this first Numerical Analysis Summer School and Workshop a successful venture.

A final word of thanks is due to Mrs. Marion Garner who handled nearly all the typing and secretarial work for the conference itself as well as some parts of this volume.

Peter R. Turner,
Department of Mathematics,
University of Lancaster.

INVITED EXPERTS

Dr. C.T.H. Baker, Department of Mathematics, University of Manchester, Manchester, England.

Prof. C. de Boor, Mathematics Research Center, University of Wisconsin, Madison, Wisconsin, U.S.A.

Prof. E.W. Cheney, Department of Mathematics, University of Texas at Austin, Austin, Texas, U.S.A.

Prof. C.W. Clenshaw, Department of Mathematics, University of Lancaster, Lancaster, England.

Dr. M.G. Cox, Division of Numerical Analysis and Computer Science, National Physical Laboratory, Teddington, Middlesex, England.

Prof. B. Engquist, Department of Mathematics, U.C.L.A., Los Angeles, California, U.S.A.

Dr. R. Fletcher, Department of Mathematics, University of Dundee, Dundee, Scotland.

Prof. P.E. Gill, Systems Optimization Laboratory, Department of Operations Research, Stanford University, Stanford, California, U.S.A.

Dr. D. Kershaw, Department of Mathematics, University of Lancaster, Lancaster, England.

Prof. K.W. Morton, Department of Mathematics, University of Reading, Reading, England.

Prof. B. Noble, Mathematics Research Center, University of Wisconsin, Madison, Wisconsin, U.S.A.

Prof. F.W.J. Olver, Institute for Physical Science and Technology, University of Maryland, College Park, Maryland, U.S.A.

Prof. B. Parlett, Department of Mathematics, University of California, Berkeley, California, U.S.A.

Prof. M.J. Todd, School of Operations Research, Cornell University, Ithaca, New York, U.S.A.

PARTICIPANTS

Dr. J.W. Barrett, Department of Mathematics, Imperial College, London, England.

Dr. G.S.J. Bowgen, DAP Support Unit, Queen Mary College, London, England.

Dr. A.R. Davies, Department of Applied Mathematics, University College of Wales, Aberystwyth, Wales.

Dr. A. Dax, Department of Applied Mathematics and Theoretical Physics, University of Cambridge, Cambridge, England.

Dr. C. Elliott, Department of Mathematics, Imperial College, London, England.

Dr. A.C. Genz, Mathematical Institute, University of Kent, Canterbury, England.

Mr. J. Gilbert, Department of Mathematics, University of Lancaster, Lancaster, England.

Dr. T.R. Hopkins, Computing Laboratory, University of Kent, Canterbury, England.

Dr. W.A. Light, Department of Mathematics, University of Lancaster, Lancaster, England.

Dr. A.J. MacLeod, Department of Mathematics, Napier College, Edinburgh, Scotland.

Dr. P. Manneback, Department of Mathematics, Facultés Universitaires de Namur, Namur, Belgium.

Dr. G. Moore, Department of Mathematics, University of Reading, Reading, England.

Dr. C. Phillips, Department of Computer Studies, University of Hull, Hull, England.

Dr. S.C. Power, Department of Mathematics, University of Lancaster, Lancaster, England.

Mr. M.R. Razali, Faculty of Mathematical Studies, University of Southampton, Southampton, England.

Dr. L.J. Sulley, Department of Mathematics, University of Lancaster, Lancaster, England

Dr. D.A. Swayne, Department of Computing Science, University of Guelph, Guelph, Ontario, Canada.

Dr. K. Tanabe, Institute of Statistical Mathematics, Tokyo, Japan.

Dr. K.S. Thomas, Faculty of Mathematical Studies, University of Southampton, Southampton, England.

Dr. R.M. Thomas, Department of Computer Studies, University of Leeds, Leeds, England.

Dr. R. Wait, Department of Computational and Statistical Science, University of Liverpool, Liverpool, England.

Dr. G. Tunnicliffe Wilson, Department of Mathematics, University of Lancaster, Lancaster, England.

LECTURE PROGRAMME

(a) Principal Lecture Courses.

C. de Boor	Multivariate Approximation.
B. Engquist	Computational Boundary Values for Initial Value Problems.
B. Noble	Topics in Numerical Analysis related to Prolongation and Restriction Operators.
B. Parlett	Matrix Eigenvalue Calculations.

(b) Shorter Lecture Courses.

C.T.H. Baker	Integral Equations.
M.G. Cox	Practical Spline Approximation.
P.E. Gill	Numerical Optimization.
K.W. Morton	Finite Element Methods for Non-Self-Adjoint Problems.
M.J. Todd	Fixed-point Methods for Nonlinear Equations.

(c) Seminars

G.S.J. Bowgen	The DAP Unit.
E.W. Cheney	Topics in Multivariate Approximation Theory (2).
A. Dax	The Downdating Problem.
R. Fletcher	1. Nonlinear Programming with an L_1 Exact Penalty Function.
	2. Second Order Corrections for Nondifferentiable Optimization.
D. Kershaw	1. Q-D Algorithms for Certain Eigenvalue Problems.
	2. Boundary Integral Problems (3).
W.A. Light	Existence Problems for Infinite Dimensional Subspaces.
F.W.J. Olver	Error Analysis of Floating-point Arithmetic (2).
S.C. Power	Best Uniform Approximation on the Unit Circle by Analytic Functions.

CONTENTS

AN INTRODUCTION TO THE NUMERICAL TREATMENT OF VOLTERRA AND ABEL-TYPE INTEGRAL EQUATIONS.

Christopher T.H. Baker.

1. ORIENTATION

An integral equation can be described as a functional equation in which the unknown function appears as part of an integrand. The inadequacies of this definition will not concern us here, since our subject is the numerical solution of a subclass of integral equations which we shall specify below. We shall frequently make comparisons with topics in the treatment of ordinary differential equations, with which we assume the reader to be more familiar.

In any specific endeavour, the numerical analyst should adopt a broad perspective. Ideally, he should relate the application area to the theoretical analysis, to the mathematical analysis of numerical schemes, and to the construction of algorithms and automatic software. Figure 1 illustrates schematically the interaction between the various areas.

Some expansion on the meaning of some of the headings in Figure 1 is in order. Under *analytic theory* we include classical and functional analysis of the equations, the theory of ill- and well-posedness in various contexts, and inherent stability* ("conditioning"), the asymptotic theory, and, for example, possible reformulation of the problem. What constitutes elegant theory may be a subjective assessment, but, frequently, the theory of discrete schemes is at its best when it mimics the theory of the functional equation.

Under *numerical methods* we include the basic numerical techniques: primitive quadrature schemes, basic collocation methods etc. The theoretical *numerical analysis* has as a first objective results on convergence, order, and numerical stability followed by the effect of rounding error. An important feature is an assessment of the ability to model, via the numerical scheme, *relevant qualitative behaviour* of the functional equations. (These considerations give rise to such notions as A-stability, for example.) We remark, in passing, that whereas many will take as a point of doctrine the necessity of convergence in a numerical scheme, as (say) some discretization parameter tends to zero, this property is not sufficient for practical computation. The modelling of important qualitative properties can be absent in practical computations with theoretically convergence schemes. An assessment of which qualitative properties are important can only result from familiarity with the area of application.

The development of reliable and robust adaptive *software* is one aim of the

---

*Observe the possibility of confusion between the use of the word stability in the sense of the classical analyst (which relates to a form of what numerical analysts term inherent conditioning) and "numerical stability".

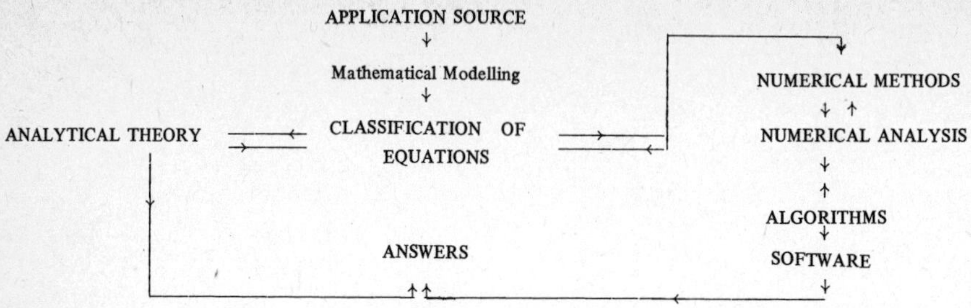

Figure 1.

practitioner: the theoretical numerical analysis frequently provides the inspiration here, but the development of *algorithms* often runs ahead of the theory. The aim is probably unrealisable without adapting one's strategy to different classes of equation, and the interactions with the theory and classification of problems are important. (Initial-value problems in differential equations may be classified, for example, as stiff or non-stiff.) In our experience, the integral equations arising in applications are all-too-frequently non-classical, and the limitations of the traditional theory are apparent. In the construction of software it seems desirable to adopt a modular construction which permits the various modules to be recombined when treating non-standard problems.

The present status of software for Volterra equations compares very unfavourably with that in other areas of interest, such as initial-value problems for ordinary differential equations. The automatic adaptive algorithms with which the author is acquainted (see [5,13] and [48], in particular) rely heavily for their motivation or implementation upon related routines for differential equations. One typical feature has already emerged in the provision of software for classical Volterra equations of the second kind: a set of test equations due to te Riele of Amsterdam has been employed to assess the properties of a number of numerical techniques [9,13,48].

Lonseth [SIAM Review 19 pp241-278, 1977] gave a survey of *application areas of* integral equations, whilst the bibliography by Noble [39] gives an extensive listing of the literature up to 1971 on integral equations and their applications. We shall be considering equations both of Volterra and Abel type, and their relative frequencies of occurrence are of interest. Whilst Abel equations certainly predominate, Volterra equations have a definite and significant rôle. We also observe that, where the analysis is feasible, Abel equations of the first kind (q.v.) may be reduced to Volterra equations of the second kind.

2. CLASSIFICATION & THEORY

In the following, the *kernel* $H(x,y,v)$ will denote a function which is assumed (unless otherwise stated) to be continuous for $-\delta \leq y \leq x+\delta \leq X+\delta$, $|v|<\infty$, for some $\delta \geq 0$, and $g(x)$ denotes a function on $[-\delta, X+\delta]$. (We may require that X can be taken arbitrarily large.) If $H(x,y,v)=K(x,y)v$ then $K(x,y)$ is called the kernel. Now,

$$\int_0^x H(x,y,f(y))dy = g(x) \qquad (0 \leq x \leq X) \tag{2.1}$$

is a *Volterra equation of the first kind* for $f(x)$. The most basic example of (2.1) is

$$\int_0^x f(y)dy = g(x) \tag{2.2}$$

where we see that the equation (in the sense of equality everywhere) has no continuous solution unless $g'(x)$ is continuous and $g(0)=0$. This reveals the possibility that (2.1) is, in a given abstract framework, likely to be ill-posed, so that admissible perturbations in $g(x)$ result in loss of existence of a solution (in a given space).

Equation (2.1) is called a *convolution equation of the first kind*, and $H(x,y,f)$ is a *convolution kernel*, if $H(x,y,f)$ has the form $H\#(x-y,f)$.

The *classical Abel equation* is the equation

$$\int_0^x \frac{1}{(x-y)^\alpha} f(y)dy = g(x) \qquad (0<\alpha<1); \tag{2.3}$$

its solution is expressible in terms of a fractional derivative of $g(x)$

$$f(x) = \frac{\sin \alpha\pi}{\pi} \frac{d}{dx} \int_0^x \frac{g(y)}{(x-y)^{1-\alpha}} dy , \tag{2.4}$$

and the equation can again be called ill-posed (though, in an imprecise way, it is less ill-posed than (2.2)).

Today, a wide class of equations is known as Abel equations (or "*equations of Abel type*"). In modern parlance

$$\int_0^x \frac{H(x,y,f(y))}{(x-y)^\alpha} dy = g(x) \qquad (0<\alpha<1) \tag{2.5}$$

is an Abel equation of the first kind. The linear case

$$\int_0^x \frac{K(x,y)}{(x-y)^\alpha} f(y)dy = g(x) \qquad (0<\alpha<1) \tag{2.6}$$

where $K(x,y)$ is continuous (at least for $0 \leq y \leq x$) is of interest.

The corresponding Volterra and Abel *equations of the second kind* are, respectively,

$$f(x) - \int_0^x H(x,y,f(y))dy = g(x) \tag{2.7}$$

and

$$f(x) - \int_0^x \frac{H(x,y,f(y))dy}{(x-y)^\alpha} = g(x). \tag{2.8}$$

Equation (2.7) is called a *convolution equation* if $H(x,y,f)$ is a convolution kernel.

The initial-value problem for an ordinary differential equation

(2.9) $$f'(x) = F(x,f(x)) \quad (x\geq 0) \quad f(0) = f_0$$

gives rise, on integrating, to a particular form of (2.7)

(2.10) $$f(x) - \int_0^x F(y,f(y))dy = f_0.$$

The equations considered above are all scalar equations but it is also possible to introduce the vector-valued counterparts such as

(2.11) $$\underset{\sim}{f}(x) - \int_0^x \underset{\sim}{h}(x,y,\underset{\sim}{f}(y))dy = \underset{\sim}{g}(x) .$$

Initial-value problems for certain *integro-differential equations* can be written in this form as a theoretical device. In particular if

(2.12) $$\begin{aligned} f'(x) &= G(x,f(x),\int_0^x H(x,y,f(y))dy) \quad (x\geq 0) \\ f(0) &= f_0 \end{aligned}$$

we obtain the system

(2.13) $$\begin{aligned} f_1(x) &= \int_0^x G(y,f_1(y), f_2(y))dy+f_0 \\ f_2(x) &= \int_0^x H(x,y,f_1(y))dy \end{aligned}$$

which can be combined to read as (2.11). The integro-differential equation is thus written as a special case of a (vector) equation of the second kind. A form of (2.12) arises on differentiating (2.7), this may be computationally useful.

The equation (2.1) can be converted, under appropriate assumptions, to a new form. Assuming the validity of differentiation, we obtain

(2.14) $$H(x,x,f(x)) + \int_0^x H_x(x,y,f(y))dy = g'(x).$$

Equation (2.14) is of a non-classical form, but in the linear case $H(x,y,f(y)) = K(x,y)f(y)$, with $K(x,x)$ non-vanishing, we find

(2.15) $$f(x) + \int_0^x \frac{K_x(x,y)}{K(x,x)} f(y)dy = \frac{g'(x)}{K(x,x)} ,$$

which is a linear equation of the second kind. Applying fractional differentiation to the linear Abel equation of the first kind

(2.16) $$\int_0^x \frac{K(x,y)f(y)}{(x-y)^\alpha} dy = g(x)$$

with the appropriate assumptions on $K(x,y)$, $g(x)$, likewise yields a Volterra equation of the second kind:

(2.17) $$f(x) + \int_0^x K\#(x,y)f(y)dy = g\#(x)$$

where

(2.18) $$K\#(x,y) = \frac{\sin \pi\alpha}{\pi}\int_0^1 \left(\frac{z}{1-z}\right)^{1-\alpha} K_x(z(x-y)+y,y)dz$$

and

(2.19) $$g\#(x) = \frac{\sin \pi\alpha}{\pi} \frac{d}{dx} \int_0^x g(y)(x-y)^{\alpha-1}dy.$$

From the above remarks it will be seen that the Volterra equation of the second kind plays a significant rôle in the analysis of classical equations and we shall emphasize (2.7) in our discussion . Non-classical equations, of which (2.14) provides one example, also arise in practice, however. By way of an example, coupled equations of the form

$$\Psi\{\theta(x)\}f(x) = 1-\int_0^x \frac{f(y)}{(x-y)^{2/3}}dy, \quad \theta(x) = \int_0^\infty k(x-y)f(y)dy$$

(where Ψ and k are prescribed functions) have recently arisen in the work of a colleague in applied mathematics.

Tsalyuk [45] provides a useful review of recent literature on Volterra equations, which includes theoretical developments up to 1976. We consider some significant results.

We have observed a relation between ordinary differential equations and integral equations and we mention here that the nonlinear Volterra equation

$$f(x) - \int_0^x \sum_{i=1}^N X_i(x)Y_i(y,f(y))dy = g(x) \tag{2.20}$$

can be solved by writing

$$f(x) = g(x) + \sum_{i=1}^N a_i(x)X_i(x) \tag{2.21}$$

where the functions $a_i(x)$ satisfy an initial-value problem for a system of differential equations [7] . This observation forms the basis for the numerical method of Bownds [13], and a similar development is possible for certain convolution equations [22]. We observe that the Volterra equation of the second kind with a polynomial convolution kernel $\Sigma\lambda_r(x-y)^r f(y)$ can be reduced to a system of differential equations [2].

More generally, the Volterra equation of second kind (2.7) can be related to a differential equation by imbedding [52]. The solution of (2.7) is $f(x)=\Psi(x,x)$ where

$$\frac{\partial\Psi}{\partial t}(x,t) = H(x,t,\Psi(t,t)), \quad \Psi(x,0) = g(x). \tag{2.22}$$

This relation is used in the study of numerical methods [49,51,52] .

The effect of perturbations in (2.7) upon the solution $f(x)$ is in theory obtainable by analysing variation-of-constants formulae [15]. In the linear case

$$f(x) - \int_0^x K(x,y)f(y)dy = g(x) \tag{2.23}$$

the variation-of-constants formula is in essence the familiar result

$$f(x) = g(x) + \int_0^x R(x,y)g(y)dy \tag{2.24}$$

where

$$R(x,y) - \int_0^x K(x,z)R(z,y)dz = K(x,y), \tag{2.25}$$

$R(x,y)=0$ if $y>x$. We can relate the *resolvent kernel* $R(x,y)$ and the *differential resolvent* $U(x,y)$;

(2.26)
$$U(x,y) = \int_y^x R(x,z)\,dz + 1$$

yields

(2.27)
$$f(x) = U(x,0)g(0) + \int_0^x U(x,y)g'(y)\,dy.$$

In classical texts the resolvent kernel is developed from the Neumann series: $R(x,y) = K(x,y) + K^2(x,y) + K^3(x,y) + \ldots$ where $K^r(x,y) = \int_0^x K(x,z)K^{r-1}(z,y)\,dz$. In the case that $K(x,y)$ is a convolution kernel, $k(x-y)$, the resolvent is a convolution kernel $r(x-y)$.

The application of variation-of-constants formula to aspects of the numerical analysis of integral equations is developed by Brunner [15] and is used also in [17].

Such formulae as those for the resolvent above are applicable to Abel equations of second kind and reveal the important fact that the solution of (2.8) may be expected to have a weak form of singularity at $x=0$ when $g(x)$ is well-behaved.

Returning to convolution kernels, we find a rôle for Laplace transforms [20,46]. If

(2.28)
$$f(x) - \lambda \int_0^x k(x-y)f(y)\,dy = g(x)$$

then, proceeding formally, presuming the transforms exist,

(2.29)
$$\mathcal{L}\{f\} = \frac{\mathcal{L}\{g\}}{1-\lambda\mathcal{L}\{k\}}$$

where

(2.30)
$$\mathcal{L}\{\phi\}(z) = \int_0^\infty \phi(t)e^{-zt}\,dt.$$

2.1. *Stability.* An aspect of the qualitative behaviour of solutions is stability.

Tsalyuk [45] includes in his survey a discussion of various stability definitions, to which we return later, and stability criteria. Such results relate to conditions on the 'input' $g(x)$ which ensure appropriate boundedness of the 'output' function, or solution, $f(x)$.

Associated with the above transform theory is the Wiener-Paley theorem, which relates to a stability result for (2.28). If $\int_0^\infty |k(t)|\,dt < \infty$ then the resolvent kernel $r(x-y)$ is such that $\int_0^\infty |r(t)|\,dt < \infty$ if and only if $\lambda\mathcal{L}\{k\}(z) \neq 1$ for $\mathrm{Re}(z) \geq 0$. The relationship with stability follows from the observation that $f(x) = g(x) + \int_0^x r(x-y)g(y)\,dy$ whence

$$\sup_{x\geq 0}|f(x)| \leq \sup_{x\geq 0}|g(x)|\{1+\int_0^\infty |r(t)|\,dt\};$$

it follows that $\sup|f(x)|$ is finite when $\sup|g(x)|$ is finite given that all the Wiener-Paley conditions are satisfied [45]. Such stability theorems are frequently stated only for the case where $\int_0^\infty |k(t)|\,dt < \infty$.

Nevanlinna considers the case where $\int_0^\infty |g(t)|^2\,dt < \infty$ and enquires [37] when $\int_0^\infty |f(t)|^2\,dt < \infty$, $f(\ )$ and $g(\ )$ being related by (2.23).

Let us proceed without undue rigour, assuming conditions additional to those stated to ensure that the solution $f(x)$ of (2.23) lies in some space F. We suppose

$\int_0^x \phi(t) \int_0^t K(t,y)\phi(y)dy\, dt$ exists and is finite for all x, and is negative for all $\phi \in F$. We shall call such a kernel *a kernel of negative type*, on F.

Writing $\|\psi\|_x$ for $\{\int_0^x |\psi(t)|^2 dt\}^{\frac{1}{2}}$, (2.23) now yields $\|f\|_x^2 \leq \|f\|_x \|g\|_x$ whence $\{\int_0^\infty |f(t)|^2 dt\}^{\frac{1}{2}} \leq \{\int_0^\infty |g(t)|^2 dt\}^{\frac{1}{2}} < \infty$. Within a suitable framework [37] Abel-type kernels of negative type can be included in such an analysis.

The Wiener-Paley result establishes conditions guaranteeing $\sup|f(x)|$ is bounded whereas in the above the results relate to $\sqrt{\int_0^\infty |f(t)|^2 dt}$. The relevance *in applications* of conditions such as $\sup|g(x)|<\infty$, the stronger condition $\int_0^\infty |g(t)|^2 dt < \infty$, and other alternatives to be found in stability definitions [45], is a subject for discussion.

Remark. The conditions which we have deduced from the Wiener-Paley theorem are not the most precise. If we consider $k(t)=1$ $(t \geq 0)$ in (2.28) then $\sup|f(x)|$ is bounded when $\sup|g(x)|$ is bounded if and only if $Re(\lambda) \leq 0$ although $\int_0^\infty |k(t)|dt$ does not here converge.

3. DISCRETIZATION METHODS

We shall consider various methods for discretizing integral equations. The methods which we describe in this section are "primitive", in the sense that we shall not be concerned with automatic adaptive strategies. We have seen that, sometimes, integral equations may be reformulated as other integral equations of a different form or as integro-differential equations-and may be approximated by ordinary differential equations-but these aspects are beyond our scope. Our first task is to describe and develop "plausible" numerical techniques but in doing so we must issue a caveat: plausible numerical techniques may have properties which make them totally or partially unsuited for certain types of equation. In particular, we have seen that first-kind equations may be regarded as ill-posed and require careful treatment: the counterparts of methods which are suited to second-kind equations may be unsuited for first-kind equations.

Discretization techniques for Volterra equations may not be suited to equations of Abel-type because of the singularity in the integrand in the latter case. Collocation methods and Galerkin methods are in principle applicable to either, but account must be taken of the possibility of (weak) singularities in the solution.

The plausible techniques for Volterra equations may be divided, roughly, into quadrature methods, classical Runge-Kutta methods, and recent modifications of these. The Runge-Kutta methods can be viewed from a number of angles, first as an extension of quadrature methods, secondly as one-step methods which are modifications of their counterparts in ordinary differential equations, and thirdly as extensions of piecewise-polynomial collocation methods.

3.1. *Quadrature rules.* For convenience, let $h>0$ be a fixed step-length. We suppose we require the approximate solution of a Volterra equation at $x \in \{0,h,2h,3h,\ldots\}$.

In the case of (2.7), we have $f(0)=g(0)$; whilst (2.1) does not yield $f(0)$ directly, the related equation (2.14) gives $H(0,0,f(0))=g'(0)$ under appropriate assumptions. Equation (2.7) yields

$$(3.1) \qquad f(ih) - \int_0^{ih} H(ih,y,f(y))dy = g(ih) \qquad (i=1,2,3\ldots)$$

and we seek methods for discretizing (3.1) in terms of $\{f(jh)\,|\,j=0,1,\ldots,i\}$ (we shall find a slightly different approach to (2.1) is permissible).

Our related quadrature problem, for example in the discretization of (3.1), is the determination of rules of the form

$$(3.2)\ Q: \qquad \int_0^{ih} \Phi(ih,y)dy \simeq h\sum_{k=0}^{i} \omega_{ik}\Phi(ih,kh) \qquad (i=1,2,3\ldots),$$

wherein we shall set, in due course,

$$\Phi(ih,y) = H(ih,y,f(y))\,;\ \omega_{ik} = 0,\ k > i.$$

Some primitive rules are provided by the "Nevanlinna rules" [37]

		k=0	k=1	k=2	...	k=i-1	k=i
(3.3)	ω_{ik} =	θ	1	1		1	$1-\theta$

wherein $\theta=1$ gives the repeated Euler rule, $\theta=0$ gives the repeated backward-Euler rule, and $\theta=\frac{1}{2}$ gives the repeated trapezium rule.

The Gregory rules, which may be regarded by some as archaic, provide useful extensions to the repeated trapezium rule: examples are [8]

		k=0	1,	2,	3,	4,	...,	i-4,	i-3,	i-2,	i-1,	i
(3.4a)	ω_{ik} =	$\frac{5}{12}$	$\frac{13}{12}$	1	1	1,	...,	1	1	1	$\frac{13}{12}$	$\frac{5}{12}$
(3.4b)	ω_{ik} =	$\frac{3}{8}$	$\frac{7}{6}$	$\frac{23}{24}$	1	1,	...,	1	1	$\frac{23}{24}$	$\frac{7}{6}$	$\frac{3}{8}$.

(There is no wisdom in employing Gregory rules of successively increasing order, without limit, as i increases.) The Gregory rules of given order cannot be employed until i is sufficiently large and we see the need for "starting procedures".

The "Kobayasi rules" are based upon the repetition of a basic rule. (Full details are in [31] and repeated in [6].) To introduce such rules, consider how the mid-point rule may be employed to generate (3.2). Our basic rule is $\int_0^{2h}\phi(y)dy\simeq 2h\phi(h)$ and for even values of i, $i=2s$, we can set

		k=0	1	2	3	4	...	i-1	i=2s
(3.5)	$\omega_{2s,k}$ =	0	2	0	2	0	...	2	0.

When i is odd ($i=2s+1$) the mid-point rule must be supplemented by another rule. We take for simplicity the trapezium rule, to obtain

		j=0	1	2	3	4	...,	i-2	i-1	i=2s+1
(3.6)	$\omega_{2s+1,k}$ =	0	2	0	2	0	...,	2	$\frac{1}{2}$	$\frac{1}{2}$.

We could consider the weights

(3.7) $\omega'_{2s+1,k} = \frac{1}{2}\,,\ \frac{1}{2}\,,\ 2\,,\ 0,\ 2\,,\ 0,..,\ 2\,,\ 0.$

instead. Other variations are possible if we use not only the trapezium rule but also Simpsons's rule and the $\frac{3}{8}$-th rule, etc. to supplement the repeated mid-point rule. One example is a rather artificial scheme with weights

(3.8)
$$\begin{array}{cccccccccc}
0 & 2 & 0 & \ldots & 2 & 0 & & & & \\
0 & 2 & 0 & & 2 & \frac{1}{2} & \frac{1}{2} & & & \\
0 & 2 & 0 & \ldots & 2 & \frac{1}{3} & \frac{4}{3} & \frac{1}{3} & & \\
0 & 2 & 0 & & 2 & \frac{3}{8} & \frac{9}{8} & \frac{9}{8} & \frac{3}{8} & \\
0 & 2 & 0 & & 2 & 0 & 2 & 0 & 2 & 0
\end{array}$$

etc.

A now classical pair of Kobayasi rules is derived from Simpson's rule (repeated) and the trapezium rule: first,

(3.9)
$$\begin{array}{lcccccccc}
 & k=0 & 1 & 2 & \ldots & 2s-1 & 2s & 2s+1 & \\
\omega_{2s,k} = & \frac{1}{3} & \frac{4}{3} & \frac{2}{3} & \ldots & \frac{4}{3} & \frac{1}{3} & & \\
\omega_{2s+1,k} = & \frac{1}{3} & \frac{4}{3} & \frac{2}{3} & \ldots & \frac{4}{3} & \frac{1}{3}+\frac{1}{2} & \frac{1}{2} & ,
\end{array}$$

and, as alternative:

(3.10)
$$\begin{array}{lcccccccc}
 & k=0 & 1 & 2 & \ldots & 2s-1 & 2s & 2s+1 & \\
\omega_{2s,k} = & \frac{1}{3} & \frac{4}{3} & \frac{2}{3} & \ldots & \frac{4}{3} & \frac{1}{3} & & \\
\omega_{2s+1,j} = & \frac{1}{2} & \frac{1}{2}+\frac{1}{3} & \frac{4}{3} & \ldots & \frac{2}{3} & \frac{4}{3} & \frac{1}{3} & .
\end{array}$$

The weights $\omega_{10} = \omega_{11} = \frac{1}{2}$ can be used to supplement the above. When used to discretize (2.7), both (3.9) and (3.10) provide convergent schemes of the same order but (3.10) provides a method which displays quite unacceptable "weak-"instability (cf.[6], Noble [38]).

The properties of integral equations derive from those of integration (as an inverse to differentiation, etc.). Not surprisingly, the properties of methods derived by discretizing via (3.2) derive from structure of the weights ω_{ij}. We can observe that all the rules Q constructed above have the following feature: the infinite array $[\omega_{ij}]$ is lower triangular and can be partitioned into the form

(3.11)

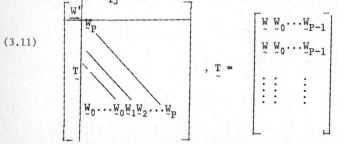

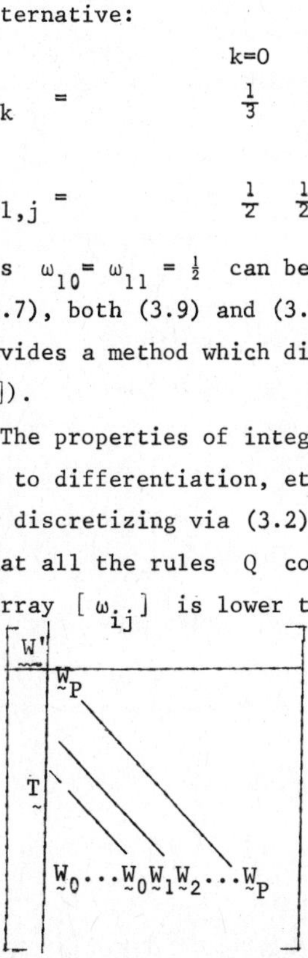

$$, \quad \underset{\sim}{T} = \begin{bmatrix} \underset{\sim}{W} & \underset{\sim}{W}_0 \cdots \underset{\sim}{W}_{P-1} \\ \underset{\sim}{W} & \underset{\sim}{W}_0 \cdots \underset{\sim}{W}_{P-1} \\ \vdots & \vdots \qquad \vdots \end{bmatrix}$$

wherein the weights of a set of starting formulae are represented by the elements of $\underset{\sim}{W}'$, and the matrices $\underset{\sim}{W}$, $\underset{\sim}{W}_0$, $\underset{\sim}{W}_1$, ..., $\underset{\sim}{W}_{P-1}$, $\underset{\sim}{W}_P$ are each of a fixed order q, say. (For the rule (3.4a), and for (3.4b), $q=1$.)

Such structure, which we generalize later, gives to the quadrature rules Q properties which can be exploited in the analysis of numerical methods. (When the weights have the form (3.11) we term them, and the rules Q, "simply-block-reducible".) The Nevanlinna rules have in addition, for appropriate values of θ, a quite different structure which Nevanlinna has exploited [37].

3.2 *Quadrature Methods* Application of the rules (3.2) to discretize (2.7) provides the system of equations

$$(3.12)\qquad \tilde{f}(ih) - h\sum_{k=0}^{i} \omega_{ik} H(ih,kh,\tilde{f}(kh)) = g(ih) \qquad (i=1,2,3,\ldots).$$

where $\tilde{f}(0) = f(0) = g(0)$. Such equations may be solved for $i=1,2,3\ldots$, in turn; for each i we have a non-linear equation (linear if $\omega_{ii}=0$) in $\tilde{f}(ih)$, to be solved by an iterative technique. As the basis of a method we may take the natural iteration

$$(3.13)\qquad \tilde{f}^{[\nu]}(ih) = \hat{g}_i + h\omega_{ii}H(ih,ih,\tilde{f}^{[\nu-1]}(ih))$$

where $\hat{g}_i = g(ih) + h\sum_{k\leq i-1} \omega_{ik} H(ih,kh,\tilde{f}(kh))$, $\tilde{f}^{[0]}(ih)$ being predicted, say, by the use of a formula similar to (3.12) in which the weights are chosen with $\omega_{ii}=0$. If $H(x,x,v)$ satisfies a Lipschitz condition in v, uniformly for $0\leq x\leq X$, then (3.13) converges for h sufficiently small. In some instances (say with larger h) a Newton iteration may be preferred.

If the equation (2.1) is treated directly, use of the rules Q provides equations

$$(3.14)\qquad h\sum_{k=0}^{i} \omega_{ik} H(ih,kh,\tilde{f}(kh)) = g(ih)$$

which differ in character from (3.12). First, if $\omega_{i0}\neq 0$ then $\tilde{f}(0)$ is required from an independent analysis. Secondly, our approach to the choice of weights may differ from that in (3.12) since the numerical scheme may not rely on all of the equations (3.14) for $i=1,2,3,\ldots$. As example, consider the use of the mid-point rule to yield for $s=1,2,3,\ldots$, using (3.5),

$$(3.15)\quad 2h\,H(2sh,(2s-1)h,\tilde{f}((2s-1)h)) = g(2sh)-2h\sum_{k=1}^{s-1} H(2sh,(2k-1)h,\tilde{f}((2k-1)h).$$

A 'starting value' $\tilde{f}(0)$ is not required. Observe that the analogous method (3.12) requires the use of the weights (3.6) in addition to (3.5) and some methods for first kind equations do not "properly" correspond to formulae for second kind equations.

In general the quadrature method for the Volterra equation of first kind yields a set of equations of the form

$$h\sum_{k=0}^{i} \omega_{ik} H(ih,kh,\tilde{f}(kh)) = g(ih), \qquad i\in\mathcal{J}\subseteq\mathbb{Z}_+ .$$

(Certain of the parameters ω_{ik} may vanish.) The linear case with $H(x,y,f(y)) = K(x,y)f(y)$ is clearly most amenable to treatment; we then see that if ω_{ii} does not vanish $(i\varepsilon\mathcal{J})$ we require $K(x,x)\neq 0$ at $x = ih$ to solve for $\tilde{f}(ih)$. A small value of $K(x,x)$ reflects a degree of ill-conditioning.

To date we have produced 'plausible' schemes only, but we have sufficient foundations to view the quadrature methods in various perspectives.

First, consider equation (2.10), the integrated form of (2.9), so that we set

$$H(x,y,v) = F(y,v), \tag{3.16}$$

in (3.12). We may examine the equations (3.12) to determine whether they are equivalent to a recognisable method for (2.9). As an example, we find that the use of the weights (3.4a) is equivalent to the application of an Adams-Moulton formula to (2.9) with starting values given by the quadrature method; (3.4b) gives the Adams-Moulton method of next higher order. To see this, we substitute (3.16) in (3.17), insert the values ω_{ik} from (3.4a) or (3.4b) in (3.12), and difference successive equations (3.12). (In this case, differencing (3.12) is an analogue of differentiating (2.10) to obtain (2.9).)

We may generalize the insight obtained here. Denote by

$$\rho(\mu) = \alpha_0\mu^m+\alpha_1\mu^{m-1}+\ldots+\alpha_m, \quad \sigma(\mu) = \beta_0\mu^m+\beta_1\mu^{m-1}+\ldots+\beta_m \tag{3.17}$$

the first and second characteristic polynomials of a linear multistep method $\{\rho,\sigma\}$ for (2.9). The equivalence (subject to variations in starting procedures) between the method $\{\rho,\sigma\}$ for (2.9) and the quadrature method (3.15) for (2.10) is preserved if the weights of the rules Q in (3.2) are $\{\rho,\sigma\}$-reducible in the sense of the following.

DEFINITION. The rules Q are $\{\rho,\sigma\}$-reducible if and only if for some $n_0>0$

$$\sum_{\ell\geq 0} \alpha_\ell\, \omega_{n-\ell,j} = \beta_{n-j} \qquad (n\geq n_0) \tag{3.18}$$

where $\omega_{kj} = 0$ for $j>k$, $\alpha_\ell,\beta_\ell = 0$ for $\ell\notin\{0,1,\ldots,m\}$.

Remark. It would be, perhaps, perverse to apply the quadrature method, as an integral equation method for (2.10), in order to solve (2.9) in practice. We now see, equally, that applying a quadrature method to (2.13) may be equivalent to applying a linear multistep method to (2.12), supplementing the latter by discretizing the integral term (written $f_2(x)$ in (2.13)) via quadrature rules Q.

Extensions of (3.18) are necessary, since the methods with weights (3.5), (3.6), (3.5), (3.7), (3.8), (3.9) and (3.10) are reducible *not* to linear multistep methods but to *cyclic linear multistep methods*, defined [43] by polynomials

$$\rho^{[\nu]}(\mu) = \sum_{\ell=0}^{m} \alpha_\ell^{[\nu]}\mu^{m-\ell}, \quad \sigma^{[\nu]}(\mu) = \sum_{\ell=0}^{m} \beta_\ell^{[\nu]}\mu^{m-\ell} \tag{3.19}$$

for $\nu=0,1,\ldots,q-1$. For completeness, we state the following:

DEFINITION. The rules Q are $\{\rho^{[\nu]},\sigma^{[\nu]}\}_0^{q-1}$ - cyclically reducible if and only if for some $n_0>0$, $n_1\geq 0$,

$$(3.20) \qquad \sum_{\ell\geq 0} \alpha_\ell^{[\nu]} \omega_{n-\ell,j} = \beta_{n-j}^{[\nu]} \qquad (n\geq n_0) \qquad \nu\in\{0,1,\ldots,q-1\}$$

where

$$n - n_1 \equiv \nu \bmod(q).$$

Given rules Q may reduce to more than one set of cyclic formulae [10].

3.2.1 The choice of quadrature rules Q in (3.2) which have been given as examples appeared as *ad hoc* constructions in terms of a basic rule. The notion of reducibility, and (3.18) and (3.20), now suggest a different avenue of approach: the construction of the weights ω_{ik} of the quadrature rules by selecting some method for (2.9), and finding some equivalent rules. Thus, choosing (3.17), we may solve (3.18) for $\{\omega_{ik}\}$. Unfortunately, the generation of the weights ω_{ik} by such a process may be inconvenient.

DEFINITIONS. (a) The weights ω_{ik} of the rules Q have (exact) repetition factor $\hat{q}$ if [32] for some j_0, j_1, n_0, $\hat{q}$ is the least positive integer such that

$$\omega_{n+\hat{q},j} = \omega_{n,j} \qquad (j_0\leq j\leq n-j_1)$$

for all $n\geq n_0$. (b) The weights have an asymptotic repetition factor $\hat{q}$ if [51]

$$\lim_{n\to\infty} (\omega_{n+\hat{q},j} - \omega_{n,j}) = 0, \; j_0\leq j\leq n-j_1.$$

If the weights have an exact repetition factor they may be computed more readily. If they have an asymptotic repetition factor then for $j_0\leq j\leq n-j$, $\omega_{n+\hat{q},j}$ and ω_{nj} agree to rounding error if n is large (with a deprecable abuse of notation "$\lim_{n\to\infty} \omega_{n+\hat{q},j} = \omega_{n,j}$") and the practical computation is again simplified. Rigorous results can be found [52] relating the existence of an exact or asymptotic repetition factor of $\{\rho,\sigma\}$-reducible rules Q to conditions upon $\rho(\mu)$.

3.3. *Runge - Kutta formulae.* If linear multistep methods are related to quadrature methods, it is natural to enquire what methods can be associated with classical Runge-Kutta methods for (2.9). Such a method is defined by a tableau

$$(3.21) \qquad \begin{array}{c|cccc} \theta_0 & A_{00} & A_{01} & \cdots & A_{0,p-1} \\ \theta_1 & A_{10} & A_{11} & \cdots & A_{1,p-1} \\ \vdots & \vdots & \vdots & & \vdots \\ \theta_{p-1} & A_{p-1,0} & A_{p-1,1} & \cdots & A_{p-1,p-1} \\ \hline \theta_p=1 & A_{p0} & A_{p1} & \cdots & A_{p,p-1} \end{array}$$

and the formulae $\tilde{f}(0) = f(0) = g(0)$,

$$(3.22) \qquad \begin{aligned} k_r^{(n)} &= hF(nh+\theta_r h, \tilde{f}(nh) + \sum_s A_{rs} k_s^{(n)}) \qquad (r=0,1,\ldots,p-1), \\ f((n+1)h) \simeq \tilde{f}((n+1)h) &= \sum_s A_{p,s} k_s^{(n)} + \tilde{f}(nh). \end{aligned}$$

If we write

(3.23) $$f_{n,r} = \tilde{f}(nh) + \sum_s A_{rs} k_s^{(n)}$$

then an equivalent formulation of the method is $\tilde{f}((n+1)h) \equiv f_{n,p}$ where

(3.24) $$f_{n,r} = f_{n-1,p} + h \sum_s A_{rs} F(nh+\theta_s h, f_{n,s}) \qquad (r=0,1,\dots,p).$$

We consider $f_{n,r}$ to be an approximation to $f(nh+\theta_r h)$. Then (3.24) may be viewed as a discretization of

$$f(nh+\theta_r h) - f(nh) = \int_{nh}^{nh+\theta_r h} F(y,f(y))dy$$

using one of the quadrature rules

(3.25) $$\int_0^{\theta_r h} \phi(y)dy \simeq h \sum_s A_{rs} \phi(\theta_s h), \qquad r=0,1,2,\dots p,$$

which are implicit in the choice of tableau (3.21). Summing (3.24) over n,

(3.26) $$f_{n,r} = h \sum_{k=0}^{n-1} \{\sum_s A_{ps} F(kh+\theta_s h, f_{k,s})\} + h \sum_s A_{rs} F(nh+\theta_s h, f_{n,s}) + f_0$$

is a discretization of (2.10). The analogous formulae for the integral equation (2.7) are

(3.27) $$f_{n,r} = h \sum_{k=0}^{n-1} \sum_s A_{ps} H(nh+\theta_r h, kh+\theta_s h, f_{k,s}) + h \sum_s A_{rs} H(nh+\theta_r h, nh+\theta_s h, f_{n,s}) + g(nh+\theta_r h), \quad r=0,1,\dots,p.$$

The method thus defined is the classical *extended Runge-Kutta method* employing the conventional Runge-Kutta array (3.21).

Remarks. (i) The Runge-Kutta array (3.21) is chosen to give high-order accuracy for (2.9). It is an interesting problem, requiring careful attention to rigour, to determine whether the same high-order accuracy results in (3.27). (ii) The array (3.21) can be modified: there seems little advantage in excluding $A_{r,p}$ $(r=0,1,\dots,p)$ from the tableau and this author replaces (3.21) by the tableau

(3.21') $$[\,\underset{\sim}{\theta} \mid \underset{\sim}{A}\,], \quad \underset{\sim}{A} \in \mathbb{R}^{p+1} \times \mathbb{R}^{p+1}$$

where $\underset{\sim}{\theta} = [\theta_0, \theta_1, \dots, \theta_p]^T, \theta_p = 1$. In this case the summation over s in the foregoing runs from $s=0$ to $s=p$. (iii) An arbitrary set of formulae (3.25) generates a tableau (3.21'); although the combination may not be optimal we can still refer to (3.27) as an extended Runge-Kutta method. *In particular*, if we write $\ell_{r,s}(\theta) = \Pi\{(\theta-\theta_t)/(\theta_s-\theta_t)\}$, the product running for $t \in \{0,1,\dots,r\}$ excluding $t=s$, we may write $A_{r,s} = \int_0^{\theta_r} \ell_{r,s}(\theta)d\theta$ $(s=0,1,\dots,r)$; $A_{rs} = 0$, $s>r$. (iv) When it is not the case that $0<\theta_0 \leq \theta_1 \leq \dots \leq \theta_{p-1} \leq \theta_p = 1$ in (3.21') the expressions in (3.27) require $H(x,y,f(y))$ with $y>x$, and such values may be undefined or unobtainable. In these circumstances we may replace the required values by those obtained by extrapolation. A number of variations are suggested by Weiss [47], Linz [32].

Runge-Kutta methods for (2.9) are commonly regarded as one-step methods in which the essential aspect of the problem is solved when one has a formula for computing $\tilde{f}(h)$ given $f(0)$. This viewpoint may be adopted in the study of

(3.28) $$f(x) - \int_0^x H(x,y,f(y))dy = g(x).$$

We may write

(3.29) $$f_n^{\#}(x) := f(x+nh), \quad g_n^{\#}(x) = \int_0^{nh} H(x+nh,y,f(y))dy + g(x+nh)$$

and then

(3.30) $$f_n^{\#}(x) = \int_0^x H(x+nh,y+nh,f_n^{\#}(y))dy + g_n^{\#}(x)$$

so that the problem of approximating $f((n+1)h)$ is that of approximating $f_n^{\#}(h)$ from (3.30). Unfortunately, $g_n^{\#}(x)$ in (3.29) is not known exactly and so it must be approximated. To deduce (3.27) in this context we write

(3.31) $$\tilde{g}_n^{\#}(t) = g(x) + h \sum_{k=0}^{n-1} \sum_s b_s H(t,kh+\theta_s h,f_{k,s}), \quad t=x+nh,$$

with the choice $b_s = A_{ps}$. However, it is equally possible to take a choice $b_s \neq A_{ps}$. Superficially it appears economical to reduce the number of terms in the sum in (3.31) and this can be achieved if we use the weights of a family of rules Q and replace (3.31) by

(3.32) $$\tilde{g}_n^{\#}(x) = g(t) + h \sum_{k=0}^{n} \omega_{nk} H(t,kh,\tilde{f}(kh)), \quad t=x+nh,$$

where $\tilde{f}(0) = f(0)$, and $\tilde{f}(h)$, $\tilde{f}(2h)$, ... are the "full-step" approximations $\{f_{n,p}\}_{n\geq 0}$ obtained from the formula. This approach yields the mixed quadrature Runge-Kutta methods:

(3.33) $$f_{n,r} = h \sum_s A_{rs} H(nh+\theta_r,nh+\theta_s h,f_{n,s})$$
$$+ g(nh+\theta_r h)+h \sum_{k=0}^{n} \omega_{nk} H(nh+\theta_r h,kh,f_{k-1,p}) \qquad (r=0,1,\ldots,p)$$

where $f_{-1,p} = f(0) = g(0)$.

3.3.1 We have developed classical Runge-Kutta formulae from the tableau (3.21). It is convenient in some respects to view the resulting Runge-Kutta methods as generalizations of the quadrature methods. To this end, we re-index the variables $\{f_{n,r}\}$ and write $\tilde{f}_0 = f(0)$ and

(3.34) $$\tilde{f}_j = f_{i,r} \quad j=i(p+1)+r+1, \quad r \in \{0,1,\ldots,p\}$$

so that $i\equiv(j-1)\mathrm{mod}(p+1)$, $r=[(j-1)/(p+1)]$ and

(3.35) $$\tau_j = ih+\theta_r h.$$

Then the classical Runge-Kutta methods developed above produce formulae of the type.

(3.36) $$\tilde{f}_j = h \sum_{k\geq 0} \Omega_{jk} H(\tau_j,\tau_k,\tilde{f}_k) + g(\tau_j).$$

thus extending the quadrature methods. The weights of the extended method are denoted

(3.37) $$\Omega_{jk} = \Omega_{jk}(\underset{\sim}{A});$$

we have $\Omega_{jk}(\underset{\sim}{A}) = \Omega_{jk}\{\underset{\sim}{b}, \underset{\sim}{A}\}$ where $\underset{\sim}{b} = [A_{p0}, A_{p1},\ldots,A_{pp}]^T$ and

$$(3.38) \qquad \Omega_{jk}\{\underset{\sim}{b},\underset{\sim}{A}\} = \begin{cases} b_t, & 0 < k \leq i(p+1) \\ A_{rt}, & i(p+1) < k \leq (i+1)(p+1) \\ 0 & \text{otherwise}; \ t \equiv (k-1) \bmod (p+1). \end{cases}$$

The weights of the mixed quadrature Runge-Kutta method are

$$(3.39) \qquad \Omega_{jk}[Q,\underset{\sim}{A}] = \begin{cases} \omega_{im}, & k=m(p+1),\ m \leq i \\ A_{rt}, & i(p+1) < k \leq (i+1)(p+1) \\ 0 & \text{otherwise}; t \equiv (k-1) \bmod (p+1). \end{cases}$$

Finally, the quadrature methods result with the choice $\Omega_{jk} = \Omega_{jk}(Q)$

$$(3.40) \qquad \Omega_{jk}(Q) = \omega_{jk}, \qquad \tau_j = jh.$$

The structure and reducibility properties of many quadrature rules have parallels in classical Runge-Kutta methods. Thus, the weights $\Omega_{jk}\{\underset{\sim}{b},\underset{\sim}{A}\}$ have the form, when partitioned

$$(3.41) \qquad \begin{matrix} \underset{\sim}{A} & & & & \\ \underset{\sim}{B}^{\dagger} & \underset{\sim}{A} & & & \\ \underset{\sim}{B}^{\dagger} & \underset{\sim}{B}^{\dagger} & \underset{\sim}{A} & & \\ \vdots & \vdots & & & \\ \underset{\sim}{B}^{\dagger} & \underset{\sim}{B}^{\dagger} & & \underset{\sim}{B}^{\dagger} & \underset{\sim}{A} \\ \vdots & \vdots & & \vdots & \end{matrix} \qquad , \ \underset{\sim}{B}^{\dagger} = \begin{bmatrix} 1 \\ 1 \\ \vdots \\ 1 \\ 1 \end{bmatrix} [b_0, b_1, \ldots, b_{p-1}, b_p] .$$

The pattern prompts the following definitions.

DEFINITION A block-lower-triangular array of weights $[\Omega_{jk}]$ is block-reducible, or $\{\underset{\sim}{A}_\ell, \underset{\sim}{B}_\ell\}_0^m$-reducible, if it may be partitioned into submatrices $\underset{\sim}{W}_{n,k}$, of order $\hat{q}$, such that (for some n_0)

$$\sum_\ell \underset{\sim}{A}_\ell \underset{\sim}{W}_{n-\ell,j} = \underset{\sim}{B}_{n-j}, \quad n \geq n_0,$$

with the conventions $\underset{\sim}{A}_\ell$, $\underset{\sim}{B}_\ell = \underset{\sim}{0}$, $\ell \notin \{0,1,\ldots,m\}$ $\underset{\sim}{W}_{n,k} = \underset{\sim}{0}$, $k>n$. If $\underset{\sim}{A}_0 = \underset{\sim}{I}$, $\underset{\sim}{A}_1 = -\underset{\sim}{I}$, $m=1$, then the weights are called simply-block-reducible.

The weights $\Omega_{jk}(\underset{\sim}{A})$ and $\Omega_{jk}\{\underset{\sim}{b},\underset{\sim}{A}\}$ are simply block-reducible. If the quadrature rules Q are cyclically reducible, $\Omega_{jk}(Q)$ are block-reducible with

$$\underset{\sim}{A}_1 = \begin{bmatrix} \alpha_q^{(0)} & \cdots & \alpha_1^{(0)} \\ \vdots & & \vdots \\ \alpha_{2q-1}^{(q-1)} & \cdots & \alpha_q^{(q-1)} \end{bmatrix}, \quad \underset{\sim}{A}_0 = \begin{bmatrix} \alpha_0^{(0)} & & \\ \alpha_1^{(1)} & \alpha_0^{(1)} & \\ \vdots & & \ddots \\ \alpha_{q-1}^{(q-1)} & \cdots & \alpha_0^{(q-1)} \end{bmatrix}$$

etc. [10]. If the rules Q are $\{\rho,\sigma\}$-reducible, or cyclically reducible, then the weights $\Omega_{jk}[Q,\underset{\sim}{A}]$ are block-reducible. Observe that (3.8) can be regarded as a modified quadrature-Runge-Kutta method when viewed appropriately.

3.3.2. The reader may object that the spirit of Runge-Kutta methods has been lost in the above development. It is the purpose in deriving Runge-Kutta formulae to take such combination of parameters θ_r, A_{rs} as to ensure that, when $g_n^{\#}(x)$ is given exactly in (3.30), the exact and computed values of $f_n^{\#}(h)$ agree to a high order in h. This direct approach has been pursued by Beltyukov, who introduces additional parameters. Considering the canonical form of (3.30), that is:

$$(3.42)\qquad \phi(x) = \int_0^x G(x,y,f(y))dy,$$

Beltyukov develops formulae of the type

$$(3.43)\qquad k_r = h\, G(\eta_r h, \theta_r h, \textstyle\sum_s A_{rs}k_s) \qquad (r=0,1,\ldots,p)$$

$$\phi(h) = \sum_s A_{ps}k_s$$

which are linked to an augmented Runge-Kutta tableau $[\ \underset{\sim}{\eta}\ |\ \underset{\sim}{\theta}\ |\ \underset{\sim}{A}\ ]$. The canonical form (3.42) can be obtained from (3.30) on deriving the integral equation for $\phi(x) = f_n^{\#}(x) - g_n^{\#}(x)$, and setting

$$G(x,y,v) = H(x+nh, y+nh, v+g_n^{\#}(x)).$$

The scheme Belyukov proposed for approximating $g_n^{\#}(x)$ is a minor variation of that used in the mixed quadrature-Runge-Kutta methods earlier.

Brunner, Nørsett and Hairer consider [16] more general schemes than Beltyukov and develop a theory, in terms of trees, to analyse the order of accuracy. A general scheme has the form

$$(3.44)\qquad f_{n,r} = h\sum_s A_{rs} H(nh+\eta_{r,s}h, nh+\Theta_{r,s}h, f_{n,s}) + \tilde{g}_n^{\#}(nh+\nu_r h)$$

wherein, with $x_{k,s} = kh + \Theta_{p,s}h$,

$$(3.45)\qquad \tilde{g}_n^{\#}(x) = g(x) + h\sum_{k=0}^{n}\sum_{s=0}^{p}\hat{\omega}_{nk}^{(s)} K(x, x_{k,s}^{(s)}, f_{k,s}),$$

the parameters

$$[\eta_{r,s}]\quad [\Theta_{r,s}]\quad [A_{rs}]\quad [\hat{\omega}_{n,k}^{(s)}]\ [\nu_r]$$

being chosen to assure high-order accuracy, inter alia. The first step of Beltyukov's method ($\tilde{g}_n^{\#}(x) = g(x)$, $n=0$) can be rewritten in this form with $\eta_{r,s} = \eta_r$, $\theta_{r,s} = \theta_s$, and (3.44) and (3.45) also includes classical methods, using (3.21), as a subset.

Finally, observe that Fehlberg-type arrays can be employed in (3.21); a somewhat different approach to estimating local errors is employed by Lomakovič and Isčuk amongst others [9 ,45].

3.4. *Special discretizations.* In developing plausible methods we have to date been following a line of thought which proceeds from (2.9) through (2.10) to (3.28). If our starting point is the discretization of (2.1) then our inspiration may be (2.2), which can be written

$$(3.46)\qquad f(x) = g'(x).$$

When we employ quadrature rules to discretize (2.1) by means of (3.15) and apply the technique to (2.2), is the resulting solution equivalent to a 'plausible' discretization of (3.46)?

DEFINITION Suppose $\{g(th)\}_{t\geq 0}$ known, $h>0$ fixed. A finite-difference method for discretizing (3.46) at $x=sh$ is called a local differentiation formula if it assumes the form

$$(3.47) \qquad \tilde{f}(sh) \simeq \frac{1}{h} \sum_k \sigma_k g((s-k)h)$$

where, for some $\kappa<\infty$, $\sigma_k = 0$ when $|k|>|\kappa|$.

Consistent formulae in which σ_k does not vanish for k sufficiently large can be constructed but their numerical stability properties are unsatisfactory.

The connection with quadrature rules can be proved by examples. We employ the mid-point rule to discretize (2.2): $2h \sum_{k=0}^{i} \tilde{f}((2k+1)h) = g((2i+2)h) \quad (i=0,1,2,\ldots)$. Differencing successive equations provides

$$\tilde{f}((2i+1)h) = \frac{1}{2h}\{g((2i+2)h) - g(2ih)\},$$

which is a local differentiation formula. Employing the repeated trapezium rule $h \sum_{k=0}^{i}{}'' \tilde{f}(kh) = g(ih)$ yields no such simplification, we find that $\tilde{f}(ih)$ depends upon $\tilde{f}(0)$ and all of $g(h),\ldots,g(ih)$.

The insight which the analysis of (2.2) provides must be interpreted with care when treating more general equations but we would certainly incline to favour the mid-point rule over the trapezium rule. (Some other rules do not even provide convergent methods.)

Taylor [44] addressing the problem of discretizing (2.1), starts with differentiation formulae (3.47) in which $\sigma_k = 0$ if $k<0$:

$$(3.48) \qquad \sigma_0\, g(sh) + \sigma_1 g((s-1)h)+\ldots+\sigma_\kappa g((s-\kappa)h) = hg'(sh)+O(h^{\nu+2}),$$

with $\nu \leq \kappa-1$, and employs (3.48) to generate 'equivalent' quadrature weights $\{\omega_{ik}\}$. Writing

$$(3.49) \qquad \sigma(\mu) = \sigma_0 + \sigma_1\mu + \ldots + \sigma_\kappa \mu^\kappa$$

the weights ω_{ik} have the form γ_{i-k} (for i,k sufficiently large) where $\gamma_0, \gamma_1, \gamma_2, \ldots$ are the coefficients of the formal power series

$$\gamma_0 + \gamma_1\mu + \gamma_2\mu^2 + \ldots = \{\sigma(\mu)\}^{-1}.$$

In a slightly earlier paper, Holyhead, McKee and Taylor analysed conditions for Kobayasi-type quadrature to be acceptable for first-kind Volterra equations: their analysis (see also Holyhead [23]) depends upon the pattern of the quadrature weights. In this context [24] it may be observed that the following concept has a rôle in the analysis:

DEFINITION. The quadrature weights $\{\omega_{ik}\}$ are said to have "*column(-wise)* repetition factor" $\hat{q}$ if $\hat{q}$ is the least integer such that for some k_0, i_0.

$$\omega_{i+\hat{q},k+\hat{q}} = \omega_{i,k}, \qquad k \geq k_0, i \geq i_0$$

Remark. The terminology is perhaps a little confusing since the repetition of weights occurs as one moves $\hat{q}$ steps along a codiagonal. In short, the existence of a column-repetition factor $\hat{q}$ ensures that the infinite array of weights $[\omega_{ik}]$ can be partitioned in the form:

$$(3.50)\qquad \left[\begin{array}{c|c} \underset{\sim}{W}' & \\ \hline \underset{\sim}{T} & \underset{\sim}{W} \end{array}\right], \qquad \underset{\sim}{W} = \begin{bmatrix} \underset{\sim}{W}^{(0)} & & & & \\ \underset{\sim}{W}^{(1)} & \underset{\sim}{W}^{(0)} & & & \\ \vdots & \ddots & \ddots & & \\ \underset{\sim}{W}^{(\nu)} & \cdots & \underset{\sim}{W}^{(1)} & \underset{\sim}{W}^{(0)} & \\ \vdots & \ddots & & \ddots & \ddots \end{bmatrix}$$

The matrix $\underset{\sim}{W}$ in (3.50), generated by $\underset{\sim}{W}^{(0)}, \underset{\sim}{W}^{(1)}, \underset{\sim}{W}^{(2)}, \ldots$, is called a block-lower-semicirculant matrix or, more concisely, *block-isoclinal* and the full matrix (3.50) is then called *bordered-block-isoclinal*. We now turn our attention to Runge-Kutta-type methods.

Perhaps the first high-order schemes for Volterra first-kind equations were those in de Hoog and Weiss [25]. The thesis of Weiss [47] contains details. Considering the linear equation

$$(3.51)\qquad \int_0^x K(x,y)f(y)dy = g(x),$$

Weiss chooses $0 \le \theta_1 < \theta_2 < \ldots < \theta_p = 1$ and derives the interpolatory quadrature formulae

$$(3.52)\qquad \int_0^{\theta_r} \phi(y)dy = \sum_{s=0}^{p} A_{rs}\phi(\theta_s) \qquad (r=0,1,\ldots,p)$$

which have degree of precision *at least* p. Two schemes result from the use of these formulae on discretizing (3.51) with $x=ih+\theta_r h$. The first assumes the form

$$(3.53)\qquad h\sum_{k\geq 0} \Omega_{jk}(\underset{\sim}{A})\, K(\tau_j,\tau_k)\, f_k = g(\tau_j)$$

where $j=i(p+1)+r+1$, $\tau_j=ih+\theta_r h$, and where $\Omega_{jk}(\underset{\sim}{A})$ are defined as in (3.37) using the coefficients A_{rs} of (3.52). (Contrast this choice with that suggested in the remark following equation (3.21').)

Because A_{rs} does not vanish for $r>s$, the preceding equations require values $K(x,y)$ with $y>x$. Accordingly, a modification may be proposed of the form:

$$(3.53)\qquad h\sum_{k\leq i(p+1)} \Omega_{jk}(\underset{\sim}{A})K(\tau_j,\tau_k)f_k + h\sum_{s=0}^{p}\theta_r A_{ps}K(\tau_j, ih+\theta_r\theta_s h)\sum_{t=0}^{p}\ell_{p,t}(\theta_r\theta_s h)f_{i(p+1)+t+1} = g(\tau_j).$$

In this expression, the second term on the left arises from the use of the quadrature rule

$$\int_0^{\theta_r}\phi(y)\,dy \simeq \theta_r\sum A_{ps}\phi(\theta_r\theta_s h),$$

and the required value $\tilde{f}(ih+\theta_r\theta_s h)$ is obtained with interpolation on approximations to the values $f(ih+\theta_0 h)$, $f(ih+\theta_1 h), \ldots, f(ih+\theta_p h)$, using the cardinal interpolation polynomials

$$\ell_{p,t} = \prod_{\substack{s \neq t \\ s=0}}^{p} \{(\theta-\theta_t)/(\theta_s-\theta_t)\}.$$

Differences between the choice $\theta_0 = 0$ and $\theta_0 > 0$ relect corresponding differences encountered earlier between the trapezium rule method and the mid-point rule method, and the choice $\theta_0 > 0$ may be preferred: then the Radau zeros may be favoured. The equations (3.52) and (3.53) are solved in a block-by-block manner, determining $f_{i(p+1)+r+1}$ for $r=0,1,\ldots,p$ simultaneously.

Keech [28,29] has developed a number of explicit methods of Runge-Kutta type which assume the form (3.53) but employ a lower-triangular matrix $\underset{\sim}{A}$ in place of that generated by (3.52), to solve first kind equations.

4. PRODUCT INTEGRATION, ETC.

The numerical techniques described above for the discretization of Volterra equations would in general fail to provide computational schemes if applied directly to the *Abel equations* (2.5) and (2.8), viz.:

$$\int_0^x (x-y)^{-\alpha} H(x,y,f(y))\,dy = g(x) \tag{4.1}$$

and $\qquad (0<\alpha<1)$

$$f(x) - \int_0^x (x-y))^{-\alpha} H(x,y,f(y))\,dy = g(x). \tag{4.2}$$

The linear versions

$$\int_0^x (x-y)^{-\alpha} K(x,y) f(y)\,dy = g(x) \tag{4.3}$$

and

$$f(x) - \int_0^x (x-y)^{-\alpha} K(x,y) f(y)\,dy = g(x) \tag{4.4}$$

have received particular attention in the literature.

Remarks. One primitive approach, *assuming that* $H(x,y,f(y))$ *is smooth as* $y \to x$, involves rewriting (4.2) in the form

$$f(x) - H(x,x,f(x)) \int_0^x (x-y)^{-\alpha} dy - \int_0^x (x-y)^{-\alpha} \{H(x,y,f(y)) - H(x,x,f(x)\}\,dy = g(x)\,.$$

The first integral is $x^{1-\alpha}/(1-\alpha)$ and the second integral now has a finite integrand and can be discretized (say) by the use of the quadrature rules Q of section 3. Solutions of Abel equations frequently demonstrate bad behaviour near $x=0$, when $g(x)$ is smooth. Where it is then possible to determine analytically the dominant asymptotic behaviour of $f(x)$ near $x=0$, this may be exploited. Thus if we establish that $f(x) = f_1(x) + f_2(x)$ where $f_2(x)$ has bounded derivative and $f_1(x) = Ax^{\beta}$, $0<\beta<1$ being known from the theory, we may be able to construct equations for $f_2(x)$ and for A.

Direct discretization of (4.1) and (4.2) requires product integration rules of the form, say,

$$(4.5)\qquad \int_0^x (x-y)^{-\alpha}\phi(y)dy \simeq \sum_k \nu_k(x)\phi(kh),$$

wherein $x\varepsilon\{ih\}_{i\geq 0}$. We may broaden the scope of our discussion if we consider Volterra or Abel equations of the form

$$(4.6a)\qquad \int_0^x H_1(x,y)H_2(x,y,f(y))dy = g(x)$$

$$(4.6b)\qquad f(x) - \int_0^x H_1(x,y)H_2(x,y,f(y))dy = g(x)$$

where $H_2(x,y,v)$ is smooth and $H_1(x,y)$ may be smooth or may have a weak singularity. Thus, some typical cases are summarized in Figure 2.

Case	Forms of $H_1(x,y)$	Case	Forms of $H_2(x,y,v)$
(i)	$K(x,y)$	(I)	(a) v or (b) $\phi(v)$
(ii)	$k(x-y)$	(II)	(a) v or (b) $\Phi(v)$
(iii)	$1/(x-y)^\alpha$ $0<\alpha<1$	(III)	$H(x,y,v)$
(iv)	1	(IV)	$H^{\#}(x-y,v)$

Figure 2.

In Figure 2, $K(x,y)$ and $k(x-y)$ are permitted to be continuous or weakly singular.

The last cases ((iv),(III),(IV)), in particular, return us to the Volterra equations considered above. The problem at hand concerns extending some of our earlier techniques. We now require, generalizing (4.5), suitable formulae

$$(4.7)\qquad \int_0^x H_1(x,y)\phi(y)dy \simeq \sum_k \nu_k(x)\phi(kh),\quad x\varepsilon\{ih\}.$$

Those quadrature formulae Q developed for use with Volterra equations, in examples in section 3, arose in general from interpolatory quadrature rules. In considering (4.7) we may choose $\{\nu_k(ih)\}$ so that

$$(4.8)\qquad \sum_k \nu_k(x)\Phi(x,y) = \int_0^x H_1(x,y)\tilde{\Phi}(x,y)dy \qquad x\ \varepsilon\ \{ih\}$$

where $\tilde{\Phi}(ih,y)$ is a piecewise polynomial interpolant to $\Phi(ih,y)$ agreeing with $\Phi(ih,y)$ for (say) $y=0,h,2h,\ldots,ih$. Examples will illustrate some generalizations of the Kobayasi rules:

Examples: We require some preliminary notation. For $k=0,1,2,\ldots$ let

$$\tilde{\Phi}_0(x,y) = \Phi(x,(2k+1)h),\quad 2kh\leq y<2(k+1)h\ ,$$

let

$$\tilde{\Phi}_1(x,y) = \{(y-kh)\Phi(x,(k+1)h)+((k+1)h-y)\Phi(x,kh)\}/h,\qquad kh\leq y\leq(k+1)h;$$

and for $k = 0,2,4,\ldots$ let

$$\tilde{\Phi}_2(x,y) = \{\tfrac{1}{2}(y-kh)(y-(k+1)h)\Phi(x,(k+2)h) + \tfrac{1}{2}(y-(k+1)h)(y-(k+2)h)\Phi(x,kh)$$
$$-(y-(k+2)h)(y-kh)\Phi(x,(k+1)h)\}/h^2,\quad kh\leq y\leq(k+2)h.$$

Then (a) the generalized repeated trapezium rule is obtained as

$$(4.9)\qquad \sum_{k=0}^{i} \nu_k(ih)\Phi(ih,kh) = \int_0^{ih} H_1(x,y)\hat{\Phi}_1(ih,y)dy.$$

(b) The generalization of (3.5) is

$$(4.10)\qquad \sum_{k=0}^{s-1} \nu_{2k+1}(2sh)\Phi(2sh,(2k+1)h) = \int_0^{2sh} H_1(2sh,y)\hat{\Phi}_0(2sh,y)dy$$

and that of (3.6) is now clear. To generalize (3.9) we write

$$(4.11)\qquad \sum_{k=0}^{2s} \nu_k(2sh)\Phi(2sh,kh) = \int_0^{2sh} H_1(2sh,y)\hat{\Phi}_2(2sh,y)dy$$

and

$$(4.12)\qquad \sum_{k=0}^{2s+1} \nu_k((2s+1)h)\Phi((2s+1)h,kh) = \int_0^{2sh} H_1((2s+1)h,y)\hat{\Phi}_2((2s+1)h,y)dy + \int_{2sh}^{(2s+1)h} H_1((2s+1)h,y)\hat{\Phi}_1((2s+1)h,y)dy.$$

The above examples illustrate how a generalization of a subset of Kobayasi rules may be constructed. As an exercise, the reader may seek a generalization of the Gregory rules.

The quadrature rules may be used to discretize, for example, equation (4.7):

$$(4.13)\qquad \tilde{f}(ih) - \sum_{k=0}^{i} \nu_k(ih)H_2(ih,kh,\tilde{f}(kh)) = g(ih)$$

for $i=0,2,3...$ where $\tilde{f}(0) = g(0)$.

Remarks. Since $f(x)$ is likely to be badly behaved at $x=0$ it is frequently advisable to employ a non-uniform step h, reducing h near to $x=0$. The required modifications will be transparent.

By a natural extension we shall generalize some of the rules employed by Weiss for Volterra equations. For convenience of presentation suppose that $0<\theta_0<\theta_1<...<\theta_p=1$ and consider a uniform stepsize h and take

$$(4.14)\qquad x \in \{\tau_j = ih+\theta_r h,\ j=i(p+1)+r+1 \geq 1\}.$$

Denote by $\hat{\Phi}_p(x,y)$ the piecewise-continuous function which is continuous on each interval $(kh,(k+1)h]$ wherein it is a polynomial of degree p in y and agrees with $\Phi(x,y)$ at $y=kh+\theta_r h$, $r=0,1,...,p$. Further, denote by $\tilde{\Phi}_j(x,y)$ the polynomial of degree r interpolating $\Phi(x,y)$ at $\{ih+\theta_s h|s=0,1,...,r\}$. Then we may construct rules

$$(4.15)\qquad \sum_{k\geq 0} \nu_k(\tau_j)\Phi(\tau_j,\tau_k) = \int_0^{ih} H_1(\tau_j,y)\hat{\Phi}_p(\tau_j,y)dy + \int_{ih}^{\tau_j} H_1(\tau_j,y)\hat{\Phi}_p(\tau_j,y)dy.$$

One analogue of (4.13) is now

$$(4.16)\qquad \tilde{f}(\tau_j) - \sum_{k\geq 0} \nu_k(\tau_j)\, H_2(\tau_j,\tau_k,\tilde{f}(\tau_k)) = g(\tau_j),$$

and whereas (4.13) can be solved in step-by-step fashion, (4.16) are to be solved in block-by-block fashion. Here, it is assumed that $H_2(x,y,f(\ ))$ can be written down for $y>x$. The simplest alternative when this is not the case involves the replacement of the last term in (4.15) by

$$\int_{ih}^{\tau_j} H_1(\tau_j,y)\tilde{\Phi}_j(\tau_j,y)dy.$$

The resulting equations (4.16) can then be solved in step-by-step fashion.

The construction of the weights $\nu_k(x)$ occurring in the above product integration schemes is considerably simplified when $H_1(x,y)$ is a convolution kernel $H_1^{\#}(x-y)$ since the infinite array of weights $[\nu_k(ih)]$ (or $[\nu_k(\tau_j)]$, as appropriate) is block-isoclinal, of the structure of $\underset{\sim}{W}$ in (3.50). Suppose, moreover, that

(4.17) $$H_2(x,y,f) = k^{\#}(x-y)f(y)$$

in (4.13) and (4.16). Then the equations (4.13) and (4.16) become linear equations in which the coefficient matrix is block-isoclinal. This observation is true, of course, when $H_1(x,y) \equiv 1$ and, for example, (4.13) become equations from a quadrature method. Note that, in the case $H_1(x,y) = (x-y)^{-\alpha}$, we find $\nu_i(ih) \propto h^{1-\alpha}$ in (4.13), and convergence of the natural iteration analogous to (3.13) occurs when $H_2(x,y,v)$ satisfies a condition $|H_2(x,x,v_1)-H_2(x,x,v_2)| \leq L|v_1-v_2|$ provided that a rather strict condition $h^{1-\alpha}LC< 1$ is satisfied (for some C). If $\alpha \sim 1$ the iteration converges slowly.

Observe that when $H_2(x,y,v) = v$ in equation (4.16) these formulae amount to the application of a collocation method for the linear equation $f(x)-\int_0^x H_1(x,y)f(y)dy= g(x)$. Two cases can be considered of particular interst: $H_1(x,y) = (x-y)^{-\alpha}$ (a classic Abel equation) and $H_1(x,y) = K(x,y)$, smooth. We refer to [14,17]. The general collocation technique for (4.2) involves determining a function $\tilde{f}(x)$ (whose restriction to any finite interval $[0, X]$ lies in a chosen finite dimensional space) constrained to satisfy (4.2) when $x \in \{z_i\}_0^N \subset [0,X]$. Analogous methods exist for (4.1); it is possible to take $\alpha=0$, so that (4.1) and (4.2) reduce to Volterra equations.

5. THEORY AND EXTENSIONS

It is difficult to motivate extensions to the material of section 3 without first providing some theoretical insight. The motivation for constructing various new methods, in particular for Volterra equations of the second kind, is then more transparent.

5.1. *Volterra equations of the second kind.* A very general convergence result can be stated for those methods which can be expressed in the form (3.36).

THEOREM 5.1. Let $\tilde{f}_j$ $(j\geq 1)$ be determined from (3.36) with $\tilde{f}_0 = f(0)$. Suppose that (i) for some $\delta>0$, $|H(x,y,v_1)-H(x,y,v_2)| \leq L|v_1-v_2|$ for $-\delta \leq y \leq x+\delta$, $x \leq X$; (ii) $\sup|\Omega_{jk}| \leq \Omega < \infty$, and (iii)

$$\lim_{h\to 0} \sup \left| \int_0^{\tau_j} H(\tau_j,y,f(y))dy - \sum_k \Omega_{jk} H(\tau_j,\tau_k,f(y_k)) \right| = 0$$

where the supremum is taken over $\tau_j = ih+\theta_r h$ such that $0 \leq ih \leq X$. Then

$$\lim_{h\to 0} \sup_{\tau_j \in [0,X]} |\tilde{f}_j - f(\tau_j)| = 0.$$

The conditions of the above theorem illustrate that for a wide class of methods for second kind Volterra equations we can dispense with worries concerning zero-stability which are found in the treatment of differential equations. Condition (ii) amounts to the condition for zero-stability

Theorem 5.1 can be strengthened. Full details appear, for example in [6,32] for quadrature methods ($\tau_j = jh$, $\Omega_{jk} = \omega_{jk}$). The order of accuracy is that of the local truncation error entering in (iii). In practice, special starting values are used to preserve high-order accuracy. However, we have remarked that in practical computations (3.9) and (3.10) behave quite differently. Theorem 5.1 only informs us that both methods yield convergent results, as $h \to 0$.

We referred above to starting values, which can be obtained from Runge-Kutta methods for example. The analysis of the order of the error in approximations obtained from the extended Runge-Kutta method (3.27) is of interest not the least because the full-step values ($f_{n,p}$ in (3.27)) display superconvergence. Insight can be obtained most readily if we consider a conventional Runge-Kutta array ($A_{rp} = 0$, $r=0,1,\ldots,p$) usually employed in the treatment of (2.9) and restrict attention to (2.20). The exact solution of (2.20) assumes the form (2.21) where

$$a_i'(x) = Y_i(y,g(x) + \sum_k a_k(x)X_k(x)) \qquad (i=1,2,\ldots,N) \tag{5.1}$$

with $a_i(0) = 0$. The approximate solution values $f_{n,r}$ obtained on applying the method (3.27) have the form

$$f_{n,r} = \sum_{k=1}^{N} \tilde{a}_k(nh+\theta_r h)X_k(nh+\theta_r h) \tag{5.2}$$

where the values $\tilde{a}_k(nh+\theta_r h)$ [compare with (3.23)] are those which would be obtained if the Runge-Kutta method were applied to the system of differential equations (5.1). The 'full-step' values $\tilde{a}_k(nh+\theta_p h)$, $\theta_p=1$, in general have high-order accuracy and since the sum in (5.2) is finite the values $f_{n,p}$ inherit this accuracy.

Remark. This argument does not extend to non-finitely separable kernels without careful attention to *rigour:* The natural extension involves approximating the given $H(x,y,f(y))$ by a separable kernel $\sum_{k=1}^{N} X_k(x)Y_k(y,f(y))$, with error ε (for $0 \leq y \leq x \leq X$, say). Clearly $N=N(\varepsilon)$ and the solutions of (2.7) and (2.20) differ on $[0,X]$ by $O(\varepsilon)$. However, considering the Runge-Kutta solution of (5.1), the constant C in results of the form $|\tilde{a}_k(nh+\theta_p h) - a_k(nh+\theta_p h)| \leq Ch^\rho$ is a function, *inter alia,* of the order N of the system of equations (5.1), and the proof via this route cannot be completed rigorously without some attention to details. For the sceptic, numerical computations certainly verify the expected error.

Observe that the remarks above provide some limited insight into the computational method of Bownds[13].

The analysis of the error in the mixed quadrature-Runge-Kutta methods is simpler. The essential analysis involves establishing the error of the one-step Runge-Kutta method (cf. (3.30)). Since the quadrature rules Q employ only the *full-step* values in (3.33) the full analysis is then easy [9]. The quadrature rules Q can be chosen to preserve high-order accuracy. The analysis of (3.31) produces problems akin to those mentioned in the above remark.

The mixed quadrature–Runge–Kutta methods employ a family Q of quadrature rules and the stability properties of the mixed methods will vary with the choice of Q. In some respects the extended methods appear to have superior stability properties, whilst being more expensive per step. We address the question of stability.

5.1.1 In his thesis, Linz [32] conjectured that *quadrature methods* using rules Q "... with repetition factor one tend to be numerically stable, those with repetition factors greater than one numerically unstable". Professor Joad might have remarked that it depends what one means by numerically stable and unstable. Linz describes qualitatively that behaviour which he considers unstable; Wolkenfelt[50] has proposed a compatible formal definition. (The first rigorous insight into this form of stability was provided by Kobayasi[31]; cf. Noble [38], and [28], [35].)

The stability concepts referred to are generalizations of those associated with parasitic solutions and growth parameters in the numerical treatment of (2.9). Sufficient insight is obtained by considering the linear case (2.23); however, an important feature of the analysis is that it also gives insight into the treatment of non-linear problems, and for this reason we discuss the effect of perturbations $\delta(x)$ in $g(x)$ rather than the effect of $g(x)$.

It is appropriate to parametrize (2.23) by λ and consider

$$f(x) - \lambda \int_0^x K(x,y) f(y)\, dy = g(x). \tag{5.3}$$

If $g(x)$ is perturbed by (a small amount) $\delta(x)$, then the resulting change in $f(x)$ is (cf. equation (2.24)).

$$\delta(x) + \lambda \int_0^x R_\lambda(x,y)\, \delta(y)\, dy, \tag{5.4}$$

where $R_\lambda(x,y)$ is the resolvent kernel $K(x,y) + \lambda K^2(x,y) + \lambda^2 K^3(x,y) + \ldots$. Consider the corresponding perturbation made in the quadrature equations

$$\tilde{f}(ih) - \lambda h \sum_{k=0}^{i} \omega_{ik} K(ih,kh) \tilde{f}(kh) = g(ih). \tag{5.5}$$

We replace $g(ih)$ by $g(ih)+\delta(ih)$. The *dominant term in* h, in the resulting perturbation to $\tilde{f}(x)$, may have a number of components one behaving like the solution of (5.4) and others (if present) behaving like the solutions $\delta_k(x)$ of equations of the form

$$\delta(x) + \mu \int_0^x R_\mu(x,y) \delta(y)\, dy, \quad \mu = \lambda_k \neq \lambda.$$

(Such components may be present in $\tilde{f}(ih)$ with coefficient $(-1)^i$, for example.)

Remark: In the nonlinear case, we set $\lambda=1$ and $K(x,y) = (\partial/\partial v)H(x,y,v)$ in the foregoing analysis, thereby considering the linearized equation (with small $\delta(x)$) for the perturbation effect.

In the above discussion, the values $\lambda_k \neq \lambda$ are to be regarded as *spurious* if we seek to model the qualitative behaviour of (5.3), or the sensitivity to perturbations of its nonlinear analogue. The values $\lambda_k \neq \lambda$ are *unwanted* if $\mu R_\mu(x,y)$ has, for $\mu = \lambda_k$, substantially different properties from $\lambda R_\lambda(x,y)$: the safest technique is to ensure that no spurious components (acceptable or otherwise) are present. Unfortunately, the analysis is based on a valid model only for small h, as $h \to 0$. The concepts of stability involved can be called *zero-relative stability*. Wolkenfelt [50] gives a formal discussion and shows that the natural interpretation of Linz's conjecture is too strong: zero-relative stable methods (methods with no spurious λ_k) which have *no* finite repetition factor can be found. Nevertheless, Linz's conjecture *is* a useful premise in the absence of further analysis.

Remark: The method using (3.10) is not zero-relatively stable, that using (3.9) is.

The above analysis does not provide insight into an appropriate choice of h. To analyze such questions it is necessary to consider regions of stability (and regions of relative stability) and a class of test equations.

The *"basic equation"*

$$f(x) - \lambda \int_0^x f(y)dy = g(x) \tag{5.6}$$

has, despite its inherent limitations (cited in [6] and subsequently repeated elsewhere by the author)still proved a useful test equation. Equation (5.6) is, of course, the integrated form of the test equation used in some of the earliest stability analysis for (2.9) and many results for (5.6) will not appear surprising on first acquaintance.

Remark: More general equations than (5.6) can be, and have been, discussed in connection with the determination of stability criteria. The insight obtained from such analysis is frequently obscured by the complexity of the theory. In consequence, the results for (5.6) are frequently appealed to. The latter results would be of little comfort had they not led to genuine insight and indeed new methods, the latter proving useful for more realistic problems than (5.6) might suggest. We observe that the question of zero-relative stability when a method is restricted to (5.6) is quite easy to settle in the light of the remarks which follow.

In the study of initial-value problems for ordinary and partial (e.g. parabolic) differential equations, the question of the numerical stability of recurrence relations of the form

$$\underset{\sim}{\Phi}_{k+1} = \underset{\sim}{M}\,\underset{\sim}{\Phi}_k + \underset{\sim}{\gamma}_k \tag{5.7}$$

is frequently encountered. For $f'(x) = \lambda f(x)$ or its counterpart equation (5.6),

$$\underset{\sim}{M} \equiv \underset{\sim}{M}(\lambda h). \tag{5.8}$$

Some definitions for (5.7) should be familiar. The recurrence relation(5.7). is *stable* if $\rho(\underset{\sim}{M}) \leqslant 1$, where if $\rho(\underset{\sim}{M}) = 1$ $\underset{\sim}{M}$ is required to be of class M. (A matrix is of class M if [40] its eigenvalues of largest modulus are semi-simple, that is, each such eigenvalue has its algebraic and geometric multiplicities equal.) The recurrence (5.7) is *strictly stable* if $\rho(\underset{\sim}{M}) < 1$.

The definitions for (5.7) enable us to consider [10] stability definitions for recurrences of the form

(5.9) $$\sum_{\ell=0}^{m} \underset{\sim}{X}_\ell \, \underset{\sim}{\chi}_{n-\ell} = \underset{\sim}{\gamma}_n \,, \quad \underset{\sim}{X}_\ell \equiv \underset{\sim}{X}_\ell(\lambda h), \quad \underset{\sim}{X}_0 = \underset{\sim}{I} \,,$$

on studying a "block-companion matrix" of the *stability polynomial*

(5.10) $$\det \sum_{\ell=0}^{m} X_\ell \mu^{m-\ell} \quad .$$

Then, (5.9) is strictly stable if (5.10) is a Schur polynomial, that is, it has its zeros in $|\mu|<1$. Further, (5.9) is stable provided that the zeros satisfy* $|\mu|\leq 1$ and those with $|\mu|=1$ are (in an appropriate sense) semi-simple.

DEFINITION. The largest of the absolute values of the roots of the stability polynomial (5.10) is called the amplification factor of the recurrence (5.9).

Remark. The reader may consider "amplification factor" to be misleading terminology since the value may be less than unity. The amplification factor is of course the spectral radius $\rho(\underset{\sim}{M})$ where $\underset{\sim}{M}$ is a block-companion matrix for (5.10); $\underset{\sim}{M}$ is frequently called the amplification matrix of (5.7).

Definitions of relative stability relate $\rho(\underset{\sim}{M})$ to $\exp(\nu\lambda h)$ where ν is some parameter associated with the structure of the vector $\underset{\sim}{\Phi}_{k+1}$ in (5.7), and will not be pursued here; see [6], however.

Equation (5.9) is a finite term recurrence relation. In contrast, the equation (5.5) relates $\tilde{f}(ih)$ to all preceding values $\tilde{f}(kh)$ $(k=0,1,\dots,i-1)$. A similar observation holds for all the Runge-Kutta equations, but for convenience the reader may limit consideration to the classical methods associated with weights Ω_{jk} and abscissae τ_k in (3.36). Thus, for the basic equation (5.5) we are concerned with the stability of equations of the form

(5.11) $$f_j - \lambda h \sum_{k\geq 0} \Omega_{jk} f_k = g(\tau_j)$$

with a choice from (3.37) through (3.40). Because of the special form of linear equation considered, and (block-)reducibility of the weights Ω_{jk}, which we shall assume, we may reduce (5.11) to a finite term recurrence.

Example. Consider the choice (3.40) and suppose the rules Q are $\{\rho,\sigma\}$-reducible, where ρ,σ are defined by (3.17). Then a simple calculation shows that (5.3) with $K(x,y) \equiv 1$, which is the particular form of (5.11) under consideration, yields

$$\sum_\ell \alpha_\ell \tilde{f}((n-\ell)h) - \lambda h \sum_\ell \beta_\ell \tilde{f}((n-\ell)h) = \sum_\ell \alpha_\ell \, g((n-\ell)h), \qquad n\geq n_0 .$$

This is a finite term recurrence with stability polynomial $\rho(\mu) - \lambda h\sigma(\mu)$. The latter polynomial is the stability polynomial for the corresponding linear multistep method applied to $f'(x) = \lambda f(x)$.

The extension of the results of the previous example to other methods is

---

* A polynomial with zeros satisfying these conditions will be called von Neumann (or, more pedantically, semi-simple von Neumann).

immediate in the presence of block-reducibility of the weights Ω_{jk} in (5.11). Let us suppose the weights Ω_{jk} partitioned into submatrices $\underset{\sim}{W}_{n,k}$ such that $\sum \underset{\sim}{A}_{\ell} \underset{\sim}{W}_{n-\ell,j} = \underset{\sim}{B}_{n-j}$, as in the definition of block-reducibility above. Then the values f_j and $g(\tau_j)$ in (5.11) can be grouped as components of vectors $\underset{\sim}{\phi}_k$ and $\underset{\sim}{\gamma}'_k$ so that (5.11) can be written

$$(5.12) \qquad \underset{\sim}{\phi}_n - \lambda h \sum_{k=0}^{n} \underset{\sim}{W}_{n,k} \underset{\sim}{\phi}_k = \underset{\sim}{\gamma}'_n$$

Since $\underset{\sim}{W}_{n,k} = \underset{\sim}{0}$ for $k>n$, the sum in (5.12) can be written $\sum_{k\geq 0}$. Then writing $n-\ell$ for n in (5.12), applying $\underset{\sim}{A}_{\ell}$ and summing we find

$$(5.13) \qquad \sum_{\ell=0}^{m} \underset{\sim}{A}_{\ell} \underset{\sim}{\phi}_{n-\ell} - \lambda h \sum_{\ell=0}^{m} \underset{\sim}{B}_{\ell} \underset{\sim}{\phi}_{n-\ell} = \sum_{\ell=0}^{m} \underset{\sim}{A}_{\ell} \underset{\sim}{\gamma}'_{n-\ell}.$$

Applying the results for (5.9), the stability polynomial (5.10) for (5.13) is

$$(5.14) \qquad \det \left[\sum_{\ell=0}^{m} \{\underset{\sim}{A}_{\ell} - \lambda h \underset{\sim}{B}_{\ell}\} \mu^{m-\ell} \right].$$

Thus the stability analysis for block-reducible methods applied to the basic equation relies on the construction of the matrices $\underset{\sim}{A}_{\ell}$, $\underset{\sim}{B}_{\ell}$ and the analysis of the zeros of (5.14).

With regard to block-reducibility, we have observed that the weights $\Omega_{jk}[Q,\underset{\sim}{A}]$ are block-reducible if the rules Q are $\{\rho,\sigma\}$-reducible or cyclically reducible. The construction of the appropriate matrices $\{\underset{\sim}{A}_{\ell},\underset{\sim}{B}_{\ell}\}$ is technical [10] but some related results are of particular interest.

If we consider the extended Runge-Kutta method with weights $\Omega_{jk}(\underset{\sim}{A})$, we find the non-trivial zero of the stability polynomial (5.14) reduces to

$$(5.15) \qquad \mu = \hat{\mu}(\lambda h) \equiv [0,0,\ldots 0,1]\,[\underset{\sim}{I}-\lambda h\underset{\sim}{A}]^{-1}[1,1,\ldots,1]^T.$$

(In general, $\hat{\mu}(\lambda h)$ is a rational approximation to $\exp(\lambda h)$.) The expression for the amplification factor $|\hat{\mu}(\lambda h)|$ reduces in the case where $\underset{\sim}{A}$ is a 'conventional' Runge-Kutta array to the amplification factor for the Runge-Kutta differential equation method applied to $f'(x) = \lambda f(x)$. Accordingly, if we choose a conventional Runge-Kutta tableau such that $|\hat{\mu}(\lambda h)| < 1$ whenever $\mathrm{Re}(\lambda h)<0$ the extended method is A-stable in the following sense.

DEFINITION. A numerical method for Volterra equations of the second kind is A-stable if, when applied to (5.6), the computed values satisfy a finite term recurrence which is strictly stable whenever $\mathrm{Re}(\lambda h)<0$.

Recall that a conventional Runge-Kutta method for (2.9) (along with the corresponding tableau $[\underset{\sim}{\theta}|\underset{\sim}{A}]$) is called A-stable when the corresponding amplification factor is such that $|\hat{\mu}(\lambda h)|<1$ whenever $\mathrm{Re}(\lambda h)<0$. Now, A-stability of an *extended* Runge-Kutta method is associated with that of the corresponding conventional method for (2.9). However, when an A-stable Runge-Kutta tableau is used in a mixed quadrature Runge-Kutta method we *cannot* expect to preserve A-stability. For, if we consider the use of $\{\rho,\sigma\}$-reducible quadrature rules Q to generate the weights $\Omega_{jk}[Q,\underset{\sim}{A}]$, the stability polynomial can be shown [10], using the analysis outlined above, to be

(5.16) $$\mu\rho(\mu) - \lambda h\,\hat{\mu}(\lambda h)\,\sigma(\mu)$$

where $\hat{\mu}(\lambda h)$ is the value in (5.15). For a consistent method $\rho(1) = 0$ so that if $\hat{\mu}(\lambda h)$ vanishes in the left half-plane (5.16) is not a Schur polynomial at that point. This superficial remark, and the example in [10] where a mixed method with an A-stable tableau is strictly stable only when $\mathrm{Re}(\lambda h)<0$ and $|\lambda h|<2$, are sufficient to indicate that loss of A-stability is likely in the mixed methods; some deeper observations are included in [27].

The stability results above are sufficient to motivate, first, the construction of some modified mixed quadrature-Runge-Kutta methods and, secondly, some analysis of methods for first-kind equations. There is some danger, however, that the over-simplification imposed by limitations on space will lead to misconceptions, and we return to stability in section 5.1.2.

5.1.1. We shall consider in this subsection the γ-modified Runge-Kutta methods originating with van der Houwen [26]. By an appropriate choice of new parameters γ_0, $\gamma_1,\ldots\gamma_p$ (see below), it is possible to modify the mixed quadrature-Runge-Kutta methods in order to obtain A-stability when the Runge-Kutta tableau is A-stable. In the process we retain the relative economy of the mixed methods over the extended method, at the price of some loss in the order of accuracy.

We shall describe the modified methods based on the weights (3.38) and (3.39) the latter being of more interest than the former. Modifications of (mixed) Beltyukov methods can also be described [26].

To motivate the methods, return to the problem of employing the weights $\Omega_{jk}[Q,\underset{\sim}{A}]$ of (3.39) to discretize (3.28) with $x \in \{\tau_j\}$. The weights (3.39) are associated with approximations

(5.17) $$\int_0^{ih} + \int_{ih}^{\tau_j} \Phi(\tau_j,y)\,dy \simeq \sum_{k\leq i(p+1)} + \sum_{k>i(p+1)} \{h\Omega_{jk}\Phi(\tau_j,\tau_k)\}.$$

The decomposition of the intervals of integration and of summation are symbolic because underlying the approximation are (in the present case) the rules Q:

(5.18) $$\int_0^{ih} \Phi(\tau,y)\,dy \simeq h\sum_{k=0}^{i} \omega_{ik}\Phi(\tau,kh).$$

(Using (3.38) the weights Ω_{jk} $(k\leq i(p+1))$ are similarly independent of $\underset{\sim}{A}$ and depend only upon i.) When $r=p$, $\tau_j = ih+\theta_p h = (i+1)h$, and $\Omega_{(i+1)(p+1),k}$ are given in (3.39) in terms of $\{\omega_{ik}\}$ and A_{ps}, $s=0,1,\ldots,p$. However, replacing i by $(i+1)$ in (5.18) provides an alternative rule *which we have not used*, and the discrepancy

(5.19) $$\tilde{f}_{(i+1)(p+1)} - h\sum_{k\geq 0}\omega_{i+1,k}\,H((i+1)h,kh,\tilde{f}_{k(p+1)}) - g((i+1)h)$$

provides an "estimate" of the accuracy at the end of the i-th step.

Such an estimate lies at the heart of the modified quadrature-Runge-Kutta methods but is used in a dynamic fashion, the estimate at the end of the (i-1)th step influencing the i-th. We shall give the formulae in a form applicable to (3.38) and

(3.39), noting that in these cases

$$(5.20)\qquad \Omega_{jk} = \Omega_{i(p+1)+1,k} \qquad k=0,1,\ldots,i(p+1),$$

for $j=i(p+1)+r+1$, $r \in \{0,1,\ldots,p\}$. Then the modified methods require a choice of the parameterizing vector

$$(5.21)\qquad \underset{\sim}{\gamma} = [\gamma_0,\gamma_1,\ldots,\gamma_p]^T \qquad 0 \leq \gamma_r \leq 1,$$

and are defined by the formulae

$$(5.22)\qquad \tilde{f}_j^{[\gamma]} = g(\tau_j) + h\sum_{k\geq 0} \Omega_{j,k} H(\tau_j,\tau_k,\tilde{f}_k^{[\gamma]}) + \gamma_r c_i^{[\gamma]}$$

$$(5.23)\qquad c_i^{[\gamma]} = \tilde{f}_{i(p+1)}^{[\gamma]} - h\sum_{k=0}^{i(p+1)} \Omega_{i(p+1)+1,k} H(ih,\tau_k,\tilde{f}_k^{[\gamma]}) - g(ih) \qquad (j=i(p+1)+r+1).$$

(Here, $c_i^{[\gamma]}$ is the "estimate" referred to above, based upon previous computed values $\tilde{f}_k^{[\gamma]}$, $k=0,1,\ldots i(p+1)$.) With the choice $\Omega_{jk} = \Omega_{jk}[Q,\underset{\sim}{A}]$ the sum in (5.23) and the terms in (5.22) for $k \leq i(p+1)$ depend only on the rules Q and the "full-step values" $\tilde{f}_{k(p+1)}^{[\gamma]}$ $(k=0,1,\ldots,i)$.

Remarks. Since $\Omega_{jk}(\underset{\sim}{A})$ have the property (5.20), the modified extended methods can be defined as above but using (3.37); however, we then find $c_i^{[\gamma]} \equiv 0$ whatever the choice (5.21), so that no new method results. *Any* quadrature rules

$$\int_0^{ih} \phi(y)dy \simeq \sum_{k=0}^{i(p+1)} \Omega^*_{i,k}\phi(\tau_k)$$

could be employed to adapt the definition in (5.23) and thus extend the γ-modified methods.

Consider the choice $H(x,y,v) = \lambda v$ in (5.22), with $\underset{\sim}{\gamma} = [1,1,\ldots,1]^T$. It is immediate that

$$(5.24)\qquad \tilde{f}_j^{[\gamma]} = \tilde{f}_{i(p+1)}^{[\gamma]} + \lambda h \sum_{k>i(p+1)} \Omega_{jk} \tilde{f}_k^{[\gamma]}$$

in view of the relation (5.20). The γ-modified method with $\gamma_r \equiv 1$ $(r=0,1,\ldots,p)$ therefore reduces, when applied to the basic test equation, to the extended method for this equation, and a particular consequence is the *A-stability of the modified method with* $\gamma_r \equiv 1$ *when the Runge-Kutta tableau is A-stable.*

Remark. The stability properties for $\underset{\sim}{\gamma} \neq [1,1,\ldots,1]^T$ are less transparent. If the γ-modified method is applied with weights $\Omega_{jk}[Q,\underset{\sim}{A}]$ where the rules Q are $\{\rho,\sigma\}$-reducible, a stability polynomial is

$$\{\mu\rho(\mu) - \lambda h\hat{\mu}(\lambda h)\sigma(\mu)\} - \gamma^{\#}\{\rho(\mu) - \lambda h\sigma(\mu)\}$$

with

$$\gamma^{\#} \equiv \gamma^{\#}(\lambda h) = [0,0,\ldots 0,1]\,[\underset{\sim}{I} - \lambda h\underset{\sim}{A}]^{-1}\underset{\sim}{\gamma}.$$

It would be disconcerting were we to find that the modified Runge-Kutta methods are not convergent. The modified methods being relatively new, we shall endeavour to provide some insight into the convergence theory and show that rather more than consistency is required in the case $\gamma = [1,1,\dots,1]^T$, an order of accuracy being lost in general with this choice of γ. We limit our analysis to what is sufficient to convey the flavour.

We define the local truncation error as

$$t_j = f(\tau_j) - h \sum_{k\geq 0} \Omega_{jk} H(\tau_j,\tau_k,f(\tau_k)) - g(\tau_j) \tag{5.25}$$
$$- \gamma_r\{f(ih) - h \sum_{k\leq i(p+1)} \Omega_{i(p+1)+1,k} H(ih,\tau_k,f(\tau_k)) - g(ih)\},$$

(where if $i=0$ the last term is taken to be zero) and the error is

$$\varepsilon_j = f(\tau_j) - f_j^{[\gamma]}. \tag{5.26}$$

In terms of t_j, ε_j we require

$$t^*_{i+1} = \max_r |t_j|, \quad \varepsilon^*_{i+1} = \max_r |\varepsilon_j| \text{ , and } t^+_{i+1} = \max_{0\leq k\leq i} t^*_{k+1}$$

the maxima being taken over $r=0,1,\dots,p$ with $j=i(p+1)+r+1$ in the first two definitions. By consistency we mean $t^+_{N+1} \to 0$ as $h\to 0$, $Nh=X$. We shall write

$$\Delta H_{jk} = H(\tau_j,\tau_k,f(\tau_k)) - H(\tau_j,\tau_k,\tilde{f}_k^{[\gamma]})$$

and assume a Lipschitz condition for $H(x,y,v)$ in v (uniformly for x,y) which with the boundedness of $|\Omega_{jk}|$ permits us to write $|\Omega_{jk}\ \Delta H_{jk}| \leq W|\varepsilon_k|$. We shall write $W^* = (p+1)W$.

Now (5.25) and the definition of $\{\tilde{f}_j^{[\gamma]}\}$ yield

$$\varepsilon_j - h \sum_{k\geq 0} \Omega_{jk}\ \Delta H_{jk} - \gamma_r\ \delta_i = t_j \tag{5.27}$$

where Ω_{jk} satisfies (5.20) and

$$\delta_i = \varepsilon_{i(p+1)} - h \sum_{k\leq i(p+1)} \Omega_{i(p+1)+1,k}\ \Delta H_{i(p+1),k} \tag{5.28}$$

Setting $j=i(p+1)$ in (5.27) and employing (5.28) we have

$$\delta_i - \gamma_p \delta_{i-1} = \zeta_i \tag{5.29}$$

$$\zeta_i = t_{i(p+1)} + h \sum_{k\geq 0} \Omega_{i(p+1)+1,k}\ \Delta H_{i(p+1),k} \tag{5.30}$$
$$- h \sum_{k\leq i(p+1)} \Omega_{i(p+1),k}\ \Delta H_{i(p+1),k}\ .$$

To proceed we consider $\gamma_p \neq 1$ and $\gamma_p = 1$ separately. In the former case (5.29) yields $|\delta_i| \leq \max_{k\leq i} |\zeta_k|/(1-\gamma_p)$ and employing (5.27) gives $\varepsilon^*_{i+1} \leq t^*_{i+1} + hW^* \sum_{k=0}^{i+1} \varepsilon^*_k + |\delta_i|$ whilst (5.30) gives $(1-\gamma_p)|\delta_i| \leq t^+_i + 2hW \sum_{k=0}^{i+1} \varepsilon^*_k$.
Thus, $\varepsilon^*_{i+1} \leq \hat{\varepsilon}_{i+1}$ where $\hat{\varepsilon}_0 = \varepsilon^*_0$ and

$$(1-h\hat{W})\hat{\varepsilon}_{i+1} = \hat{t}_{i+1} + h\ \hat{W} \sum_{k=0}^{i} \hat{\varepsilon}_k \tag{5.31}$$

where $\hat{W} = \{1+2/(1-\gamma_p)\}W^*$, $\hat{t}_{i+1} = t^*_{i+1} + t^+_i/(1-\gamma_p) \leq t^+_{i+1}\{1+1/(1-\gamma_p)\}$.

Equation (5.31) is of a familiar type [6,p925] in the analysis of methods for Volterra equations of the second kind. It yields a convergence result from consistency, since (5.31) gives

$$\hat{\varepsilon}_i \leqslant \exp\left\{\frac{ih\hat{W}}{1-h\hat{W}}\right\} \{\varepsilon_0^* + \max_{k\leqslant i} \hat{t}_k /(1-h\hat{W})\}$$

and $h \to 0$ with $ih = x \leqslant X$.

The preceding analysis indicates the simplicity of the convergence theory with $\gamma_p \neq 1$ but does not apply with $\gamma_p = 1$. A proof for arbitrary $\underset{\sim}{\gamma}$ is supplied under modified assumptions in [27]. Since our aim is to provide additional insight we illustrate the case $\gamma_p = 1$ by considering $\underset{\sim}{\gamma} = [1,1,1,\ldots,1]^T$.

In addition to our earlier assumptions we impose the condition (5.20) and assume that the derivative $H_x(x,y,v)$ exists and satisfies a Lipschitz condition in v (uniformly in x,y). It follows that there exists an L_1 such that $|\Delta H_{jk} - \Delta H_{i(p+1),k}| \leqslant \theta_r L_1 h\, \varepsilon_k|$. In consequence, there exists a constant V such that

$$|\Omega_{jk}\Delta H_{jk}| \leqslant V|\varepsilon_k|,\ |\Omega_{jk}(\Delta H_{jk} - \Delta H_{i(p+1),k})| \leqslant hV|\varepsilon_k|.$$

Now rearranging (5.27) yields $\varepsilon_j = \varepsilon_{i(p+1)} + h\sigma_j + t_j$ where by (5.20)

$$\sigma_j = \sum_{k\leqslant i(p+1)} \{\Omega_{jk} - \Delta H_{jk} - H_{i(p+1),k}\} +$$

$$+ \sum_{k>i(p+1)} \Omega_{jk}\,\Delta H_{jk}$$

and hence

$$|\sigma_j| \leqslant Vh \sum_{k\leqslant i(p+1)} |\varepsilon_k| + V \sum_{i(p+1)+1}^{(i+1)(p+1)} |\varepsilon_k|$$

Employing our earlier notation and setting $V^* = (p+1)V$ we now find, provided all $\tau_j \in [0,X]$ and h is sufficiently small

$$\hat{e}_{i+1} = \rho_h\,\hat{e}_i + \frac{t_{i+1}^+}{1-h\,V^*} \tag{5.32}$$

where $\rho_h = (1+hV^*X)/(1-hV^*)$. Results of the form (5.32) are common in the study of initial value problems for differential equations and yield convergence when $\varepsilon_0^* \to 0$ and $t_{i+1}^+ = o(h)$ as $h \to 0$ with $ih\varepsilon[0,X)$. The condition $t_{i+1}^+ = o(1)$ is apparently not sufficient. An order of convergence appears to be lost because $\rho_h - 1$ behaves like h as $h \to 0$.

The above analysis is intended to be sufficient to provide some insight into the modified methods; precise results on the superconvergence expected at full-step values require analysis which we do not develop here for reasons of space. Further reading is available in [26,27].

5.1.2. The γ-modified Runge-Kutta methods of §5.1.1 may be motivated by stability considerations in terms of a basic test equation.

We shall devote some space, in this subsection, to reassessing (and placing in perspective) the stability analysis which we gave for certain classical methods applied to the basic equation.

First, observe that the analysis relied upon an ability to reduce the relation between successive computed values to a finite-term recurrence. Structure of the weights Ω_{jk} in (3.36) is therefore required. For (5.6) the recurrence relation is a constant-coefficient one; the amplification matrix in (5.7) is independent of k. However, somewhat similar techniques can be applied whenever the integral equation assumes the form (2.10) or the integrated form of a system of equations and the analysis for the basic equation can in consequence be modified and extended [2, 7] to polynomial convolution kernels and, for example, separable kernels (cf. (2.20)). Imbedding also provides an opportunity for exploiting the reducibility structure: for example, if the quadrature method is applied with $\{\rho,\sigma\}$-reducible rules Q then the computed values are related to the function satisfying the discrete analogue of (2.22): $\sum \alpha_\ell \tilde{\Psi}(x,(n-\ell)h) = h\sum\beta_\ell H(x,(n-\ell)h,\tilde{\Psi}((n-\ell)h,(n-\ell)h))$.

Ramifications of this result are explored in [52] and this analysis has also led to the construction of a new class of multilag and modified multilag methods [49].

There is some danger that the manipulative details of the stability analysis may obscure the underlying philosophy. To place our subject in perpective it is appropriate to study the qualitative behaviour of solutions to integral equations. It is sufficient for our discussion to consider the linear equation

$$f(x) - \int_0^x K(x,y)f(y)dy = g(x), \qquad \infty > x \geq 0. \tag{5.32}$$

The remarks centered upon (2.23) apply, and the resolvent kernel or related functions show how $f(x)$ reacts to perturbations in $g(x)$. (In particular if $g(x)$ is perturbed by a constant ε, (2.27) shows that $f(x)$ changes by $\varepsilon\, U(x,0)$.) Stability definitions for (5.32) (see [45]) require the specification of an input set $G = \{\delta g\}$ of perturbations to g and an output set F of perturbations to f. Then (5.32) is *stable* if $\delta g \in G$ implies a change δf in f with $\delta f \in F$ and $\|\delta f\|_F \leq C\|\delta g\|_G, C<\infty$, the norms being associated with the choice of sets F,G. Equation (5.32) is in addition *asymptotically stable* if it is stable and $\delta f(x) \to 0$ as $x \to \infty$.

Remark: For nonlinear equations, stability definitions refer to a particular solution $f(x)$ of the nonlinear equation.

Replacing (5.32) by discrete equations, such as the equation

$$\tilde{f}(ih) - h\sum_{k=0}^{i} \omega_{ik} K(ih,kh)\tilde{f}(kh) = g(ih) \tag{5.33}$$

obtained using quadrature rules Q, it is appropriate to enquire whether the stability properties of (5.33) reflect relevant stability properties of (5.32). Thus, perturbations $\delta g \in G$ for (5.32) induce perturbations $\delta \underset{\sim}{g} = \{\delta g(ih), i\geq 0\} \in \tilde{G}$ which may be measured in some norm $\|\cdot\|_{\tilde{G}}$, and one may ask (for an appropriate $\tilde{F}$) that $\delta\underset{\sim}{\tilde{f}} = \{\delta f(ih), i\geq 0\} \in \tilde{F}$ and $\|\delta \underset{\sim}{\tilde{f}}\|_{\tilde{F}} \leq C\,\|\delta \underset{\sim}{g}\|_{\tilde{G}}$ (stability) or in addition $\delta\tilde{f}(ih) \to 0$ as $i \to \infty$ (asymptotic stability). Nevanlinna [37] adopts this framework when considering positive quadratures.

Observe the question of modelling in the choice of $\tilde{G}$. If $\tilde{f}(ih)$ (as $i\to\infty$) is to mimic $f(x)$ as $x \to \infty$ we may be moved to ask that (5.33) is asymptotically stable in a framework corresponding to a choice of $\tilde{G}$ which contains as an element $t = \{t_i : = f(ih) - h\sum_k \omega_{ik} K(ih,kh) f(kh) - g(ih),\ i \geq 0\}$. (If rounding error is included, an alternative choice suggests itself.) However, we cannot assume that $t_i \to 0$ as $i \to \infty$ unless (5.33) is restricted so a choice $\tilde{G}$ normed by $\{\sum_0^\infty |t_i|^p\}^{1/p}$ (p=1,2), for example, is inappropriate.

The stability of (5.32) may be examined in terms of the inverse $(I - \lambda K)^{-1}$: formally, $f = (I - \lambda K)^{-1} g$; the inverse operator is implicit in (2.24). In a like manner, we consider (5.33) and denote by Z_N the $N \times N$ coefficient matrix. Suppose that $\delta\tilde{g} = [\delta g(0), \delta g(h), \delta g(2h), \ldots]^T \in \tilde{G}$, $\delta\tilde{g}^{[N]} = [\delta g(0), \delta g(h), \ldots, \delta g(Nh), 0, 0 \ldots]^T \in \tilde{G}$, $\delta\tilde{g}_N = [\delta g(0), \delta g(h), \ldots, \delta g(Nh)]^T \in \mathbb{R}^{N+1}$, with corresponding definitions for the perturbations in $\{\tilde{f}(ih)\}$. We suppose $\|\delta\tilde{g}^{[N]}\|_{\tilde{G}} = \|\delta\tilde{g}_N\|_N$ in an appropriate norm on $\mathbb{R}^{N+1}$ and we find

$$\|\delta\tilde{f}_N\|_N \leq \|Z_N^{-1}\|_N \, \|\delta\tilde{g}_N\|_N .$$

where Z_N depends upon the given $h > 0$. It follows that if $\|Z_N^{-1}\|_N$ is uniformly bounded with N,

$$\|\delta f\|_{\tilde{G}} \leq \limsup_{\substack{N\to\infty \\ h \text{ fixed}}} \{\|Z_N^{-1}\|_N\} \, \|\delta g\|_{\tilde{G}}$$

yielding a stability result where the norms for $\tilde{F}$, $\tilde{G}$ coincide. Moreover, examining the N-th row of Z_N^{-1} may permit a conclusion concerning asymptotic stability. In the above approach, the nature of the perturbations $\delta\tilde{g}$ plays a rôle.

Let us relate the foregoing to our previous stability analysis, taking $K(x,y) \equiv 1$. For simplicity we consider the quadrature method employing $\{\rho,\sigma\}$-reducible rules. It follows that for $K(x,y) \equiv 1$,

$$A_N Z_N = A_N - \lambda h B_N$$

where A_N and B_N are sparse bordered semi-circulant(isoclinal) matrices, say:

$$A_N = \left[\begin{array}{c|cccc} I & & & & \\ \hline \alpha_m \ldots \alpha_1 & \alpha_0 & & & \\ & \ddots & \ddots & \ddots & \\ & & \alpha_m \ldots \alpha_1 & & \alpha_0 \end{array}\right], \qquad B_N = \left[\begin{array}{c|cccc} Z_{n_0} & & & & \\ \hline \beta_m \ldots \beta_1 & \beta_0 & & & \\ & \ddots & \ddots & \ddots & \\ & & \beta_m \ldots \beta_1 & & \beta_0 \end{array}\right]$$

Thus the reducibility permits an elimination process to be applied to the matrix Z_N to reduce it to bordered isoclinal form. Rather than compute $Z_N^{-1} = [I_N - \lambda h A_N^{-1} B_N]^{-1}$ we resort to recurrence relations (5.7) to pursue the analysis. However, it will be observed that the resulting vectors γ_k in (5.7) have a form which depends upon the nature of δg. Whilst it is commonplace to consider stability definitions for (5.7) which depend *only* upon the amplification matrix M, the nature of the vectors γ_k has a rôle in the present context. Indeed, consider the question of asymptotic stability and (5.7). If $\rho \equiv \rho(M) < 1$ then

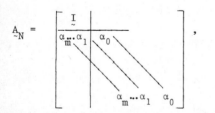

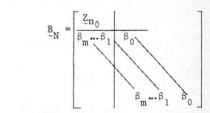

$\| \Phi_{k+1} \| \leq \rho \| \Phi_k \| + \| \gamma_k \|$ and $\| \Phi_k \| \leq \| \gamma_0 \| \rho^k + \| \gamma_1 \| \rho^{k-1} + \ldots \| \gamma_{k-1} \|$.
We ask that $\| \Phi_k \| \to 0$ as $k \to \infty$, when considering asymptotic stability. This is achieved when $\gamma_1 = \gamma_2 = \ldots = \gamma_k = 0$ in (5.7) and in particular when $\delta g(ih)$ is constant for all i and $\sum \alpha_\ell \delta g((n-\ell)h) = 0$; the latter relation then follows from $\rho(1) = 0$.

The foregoing remarks suggest an alternative definition of A-stability: "A numerical method for equations of the second kind is A-stable if the computed values decay to zero when the method is applied to the basic equation with $Re(\lambda h) < 0$ and $g(x) \equiv 1$."

Remark: The discussion for Runge-Kutta methods differs slightly from the above if we are concerned only with the full-step approximations.

If one endeavours to extend the above analysis via reducibility it is quite appropriate to consider block-reducibility of a matrix of weights where the matrices A_ℓ, B_ℓ *vary* with h; $A_\ell = A_\ell(h)$, $B_\ell = B_\ell(h)$. Thus, for the product integration methods the weights $\nu_k(ih)$ are unlikely to have the simple structure we have exploited unless $K(x,y)$ satisfies a difference equation. Indeed, the matrix Z_N arising from structured quadrature methods with $K(x,y) \not\equiv 1$ is already less amenable to treatment. In the important case of convolution kernels, however, the matrix Z_N is (given structured quadrature) bordered-block-isoclinal and the author has recently developed the basis of a technique for discussing such cases. Product integration techniques for Abel equations also yield matrices of this form.

It is appropriate, before proceeding, to constrast the stability discussion above with the concept of zero-stability, which would require that $\limsup_{\substack{N\to\infty \\ Nh=X}} \|Z_N^{-1}\|_\infty < \infty$.

5.2. *Abel equations, etc.* Volterra equations of the second kind are more amenable to analysis than those of the first kind and Abel equations.

With regard to Volterra equations of the first kind we note that for methods which are counterparts of second kind methods, a connection between the stability analysis for the basic equation of the second kind can be established. Such methods correspond to local differentiation formulae when the recurrence relation for the basic equation reduces to the form (5.7) and $\rho(M(\lambda h)) \to 0$ as $Re(\lambda h) \to -\infty$. (For related remarks see [28] and [51]. This condition is stronger than A-stability [8].) Thus, it would appear that L_0-stable linear multistep methods $\{\rho, \sigma\}$ might generate appropriate rules Q for quadrature methods for first kind equations: see Wolkenfelt [51] for precise results along these lines.

Convergence results for linear Volterra equations of the first kind (3.51) are frequently obtained under the assumption $|K(x,x)| \geq \alpha > 0$. The case where this restriction is not valid is of interest.

Convergence results for product integration methods applied to certain equations of the second kind, were given by Linz [32]. Recently, Kershaw [30] has shown, in the case of Abel equations, the rôle in the convergence analysis of the Mittag-Leffler function.

There is no direct connection between Abel equations of the second kind and initial-value problems in differential equations and the more simple theories in Volterra equations cannot readily be mimicked in the treatment of Abel equations. Brunner and Nørsett, in a useful paper [17], treat collocation methods for Abel and Volterra equations in a unified framework; they contrast the results obtainable for Volterra equations of the second kind with the less favourable ones obtainable with Abel equations.

The stability theory for numerical methods for Abel equations of the second kind is relatively in its infancy and reflects corresponding problems in the parallel theory of the functional equation.

For Abel equations of the first kind, Eggermont [21] has produced a pleasing analysis of some primitive product integration methods for Abel equations of the *first kind*. His analysis mimics the reduction of (2.16) to (2.17) and permits fairly precise statements concerning the error in the approximate solution. Eggermont's paper [21] gives as references earlier work related to his study. The work of Cameron [18] is also, we believe, related.

Brunner [communicated privately] has recently surveyed the literature on Volterra and Abel integral equations. As will be apparent from the emphasis here, and from consulting that survey, there remain a number of areas of interest for further work - particularly concerning Abel equations and Volterra equations of the first kind. By concentrating upon Volterra equations of the second kind we address a tractable problem in which the insight obtained moulds our expectations of what might be achieved for more difficult problems.

6. REFERENCES.

1. AMINI,S.Stability analysis of methods employing reducible quadrature rules for solutions of Volterra integral equations. Report CS-81-02, School of Math., University of Bristol (1981).

2. AMINI,S.,BAKER, C.T.H., van der HOUWEN, R.J. & WOLKENFELT, P.H.M. Stability analysis of numerical methods for Volterra integral equations with polynomial convolution kernels. Rep. NW 109/81 Math Centrum, Amsterdam (1981).

3. ANDERSSEN,R.S,de HOOG, F.R., & LUKAS, M.A. Application and numerical solution of integral equations, Sijthoff & Noordhoff, Alphen aan de Rijn (1980).

4. ANDRADE,C.& McKEE, S. On optimal high accuracy linear multistep methods for first kind Volterra integral equations. BIT 19 pp.1-11 (1979).

5. APPELBAUM, L. and BOWNDS, J.M. A FORTRAN subroutine for solving Volterra integral equations, Math. Rept., Univ. of Airzona, Tucson (1981).

6. BAKER, C.T.H. The numerical treatment of integral equations, Clarendon Press, Oxford (1977, reprinted 1978).

7. BAKER, C.T.H. Structure of recurrence relations in the study of stability in the numerical treatment of Volterra integra and integro-differential equations. J.Int. Eqns. 2 pp.11-39 (1980).

8. BAKER, C.T.H. & KEECH, M.S. Stability regions in the numerical treatment of Volterra integral equations. SIAM J.Numer.Anal. 15 pp. 394-417 (1978).

9. BAKER, C.T.H., RIDDELL, I.J., KEECH, M.S., & AMINI, S. Runge-Kutta methods with error estimates for Volterra integral equations of the second kind. ISNM 53 pp.24-42 (1980).

10.BAKER,C.T.H. & WILKINSON,J.C. Stability analysis of Runge-Kutta methods applied to a basic Volterra integral equation. J. Austral. Math. Soc(B) 22 pp.515-538 (1981).

11.BENSON,M.P.Errors in numerical quadrature for certain singular integrands and the numerical solution of Abel integral equations. Ph.D. thesis, Univ. of Wisconsin-Madison, 1973 (Univ. Microfilm. 74 - 3,509).

12.BOWNDS, J.M. On an initial-value method for quickly solving Volterra integral equations - a review. J.Opt. Theory. Appl.24 (1978) pp. 133-151.

13.BOWNDS, J.M. Theory and performance of a subroutine for solving Volterra integral equations. Math. Rept., Univ. of Arizona, Tucson (1981).

14.BRUNNER, H. On superconvergence in collocation methods for Abel integral equations. Proc. 8th Manitoba Conf. pp. 117-128, Manitoba (Winnipeg) (1978).

15.BRUNNER, H. The application of the variation of constans formulas in the numerical analysis of integral and integro-differential equations. Math. Rep. No. 20, Dalhousie Univ. (& Utilitas Math. 19 pp.255-290 (1981)).

16.BRUNNER, H., HAIRER, E. & NØRSETT, S. P. Runge-Kutta theory for Volterra integral equations of the second kind. Math. Comp.(to appear).

17.BRUNNER, H. & NØRSETT, S.P. Superconvergence of collocation methods for Volterra and Abel integral equations of the second kind. Numer. Math. 36 pp.347-358 (1981).

18.CAMERON, R.F. Direct solution of Volterra integral equations. D.Phil. thesis, Oxford Univ. (1981).

19.CAMERON, R. F. & McKEE, S. High accuracy convergent product integration methods for the generalized Abel equation. Ms., Oxford Univ. Comp. Lab. (1981).

20.DOETSCH, G. Introduction to the theory and application of the Laplace transformation. Springer, New York (1978).

21.EGGERMONT, P.P.B. A new analysis of the Euler-midpoint-and trapezoidal-discretization methods for the numerical solution of Abel-type integral equations. Dept. Comp. Sci. Tech. Rep., SUNY Buffalo. (J.Int.Eqn. to appear).

22.GOLBERG, M.A. (ed.) Solution methods for integral equations: theory and applications, Plenum Press, New York (1979).

23.HOLYHEAD, P.A.W. Direct methods for the numerical solution of Volterra integral equations of the first kind. Ph.D. thesis, Univ. of Southampton (1976).

24.HOLYHEAD, P.A.W. & McKEE, S. Stability and convergence of multistep methods for linear Volterra equations of the first kind. SIAM. J.Numer.Anal.13 pp.269-292 (1976).

25.de HOOG F. & WEISS, R. On the solution of Volterra integral equations of the first kind. Numer. Math. 21 pp. 22-32 (1973).

26.van der HOUWEN, P.J. Convergence and stability analysis of Runge-Kutta type methods for Volterra integral equations of the second kind. Rep. N.W.83/80 Math. Centrum, Amsterdam (1980) and BIT 20 pp. 375-377.

27.van der HOUWEN, P.J., WOLKENFELT, P.H.M. & BAKER, C.T.H. Convergence and stability analysis for modified Runge-Kutta methods in the numerical treatment of second-kind Volterra integral equations. IMA J.Numer.Anal. 1 pp.303-328 (1981).

28.KEECH, M.S. Stability in the numerical solution of initial-value problems in integral equations. Ph.D. thesis, Univ. Manchester (1977).

29.KEECH, M.S. A third-order semi-explicit method in the numerical solution of first kind Volterra integral equations. BIT 17 pp. 312-320 (1977).

30.KERSHAW, D. Lecture notes: seminar , Imperial College, London, U.K. (17th Nov.1980).

31.KOBAYASI, M. On the numerical solution of the Volterra integral equations of the second kind by linear multistep methods. Rep Statist. Appl. Res. Un. Japan Sci.Engrs. 13 pp. 1-2- (1966).

32.LINZ, P. The numerical solution of Volterra integral equations by finite difference methods.MRC Tech. Rep. 825, Univ. of Wisconsin - Madison (1967).

33.LINZ.P. Numerical methods for Volterra integral equations with singular kernels. SIAM J. Numer. Anal. 6 pp. 365-374 (1969).

34.LOGAN, J.E. The approximate solution of Volterra integral equations of the second kind. Ph.D. thesis Univ. Iowa (1976).

35.McKEE, S. & BRUNNER, H. The repetition factor and numerical stability of Volterra integral equations. Comput.Math.Appl. 6 pp.339-347 (1980).

36.MILLER, R.K. Nonlinear Volterra integral equations. Benjamin, Menlo Pk. (1971).

37.NEVANLINNA, O. Positive quadratures for Volterra equations. Computing 16 pp.349-357 (1976).

38.NOBLE, B. Instability when solving Volterra integral equations of the second kind by multistep methods. Springer Lect.Notes.Math. 109 pp. 23-39 (1969).

39.NOBLE, B. A bibliography on methods for solving integral equations. MRC Repts. 1176 & 1177 Univ. of Wisconsin at Madison (1971).

40.ORTEGA, J.M. Numerical analysis: a second course. Academic, New York, 1973.

41.te Riele, H.J.J. (ed.) Colloquium:numerical treatment of integral equations·MC syllabus No.41., Math. Centrum, Amsterdam (1979).

42.te RIELE, H.J.J. Collocation methods for weakly singular second kind Volterra integral equations with non-smooth solution. Ms., Math. Centrum, Amsterdam (1981).

43.STETTER, H. Analysis of discretization methods for ordinary differential equations. Springer, New York (1973).

44.TAYLOR, P. The solution of Volterra integral equations of the first kind using inverted differentiation formulae. BIT 16 pp. 416-425 (1976).

45.TSALYUK,Z.B.Volterra integral equations. J. Soviet. Math. 12 pp. 715-758, 1979.

46.WATSON, E.J. Laplace transforms and applications. van Nostrand, New York, 1981.

47.WEISS, R. Numerical procedures for Volterra integral equations. Ph.D. thesis, ANU (Canberra) (1972).

48.WILLIAMS, H.M. Variable step-size predictor-corrector schemes for Volterra second kind integral equations. D. Phil. Thesis, Univ. Oxford (1981).

49.WOLKENFELT, P.H.M. Modified multilag methods for Volterra functional equations. Rep. NW108/81, Math. Centrum, Amsterdam (1981). See [51] also.

50.WOLKENFELT, P.H.M. On the relation between the repetition factor and numerical stability of direct quadrature methods for second kind Volterra integral equations. Rep, NW /81 Math. Centrum, Amsterdam (1981). See [51] also.

51.WOLKENFELT, P.H.M. The numerical analysis of reducible quadrature methods for Volterra integral and integro-differential equations. Academisch Proefsdinft, Math. Centrum & Univ. Amsterdam (1981).

52.WOLKENFELT, P.H.M., van der HOUWEN, P.J. & BAKER, C.T.H. Analysis of numerical methods for second kind Volterra equations by imbedding techniques. J.Int. Eqns. 3 pp. 61-82 (1981).

Department of Mathematics,
University of Manchester,
Manchester.

Topics in Multivariate Approximation Theory

C. de Boor
Mathematics Research Center
University of Wisconsin
Madison WI 53706

Outline

1. Approximation Theory

In this first lecture, I intend to give an overview of what is understood by the term "Approximation Theory". This is both a bow toward the title of these lectures and a survey of the kinds of things you might reasonably expect to see covered in these lectures, albeit with the special accent of "multivariate", but which I will for the most part not cover at all. In effect, this allows you to locate within the large scheme of things the few specific items I do cover.

Approximation Theory is usually understood to deal first and foremost with **best approximation**, or **b.a.** for short. This is the task of finding, given an element x of some **metric space** X, an element m^* from some given subset M of X for which

$$\mathrm{dist}(x,m^*) = \inf_{m\in M} \mathrm{dist}(x,m) =: \mathrm{dist}(x,M) .$$

Such an m^* is called a **b.a. to x from M**. In symbols:

$$m^* \in \mathcal{P}_M(x) .$$

Basic questions asked concern:

Existence: $|\mathcal{P}_M(x)| > 0$?

Uniqueness: $|\mathcal{P}_M(x)| < 2$? More generally, $|\mathcal{P}_M(x)| = $?

Characterization: How would one recognize a b.a. (other than by the brute force approach of comparing it with all candidates)? This is particularly important for the next question.

Construction.

A priori bounds: What can be said about $\mathrm{dist}(x,M)$ based on the information that x lies in some set K ?

Details of the answers depend strongly on the specifics of X, dist, and M. Most commonly, X is a normed linear space, such as

$C(T)$:= continuous functions on some locally compact metric space T

(e.g., $T = [a,b]$ or $T = \mathbb{R}^n$ or whatever) and the metric is provided by the norm on X.

Existence requires that M be closed. Beyond that, it is usually a matter of local compactness: A **minimizing sequence** (m_n) in M is picked; this means that $m_n \in M$ and

$$\lim_{n\to\infty} \|x - m_n\| = \mathrm{dist}(x,M) .$$

Then $\{m_n : n=1,2,\ldots\}$ is bounded, hence, by local compactness, has a limit point m in M. For this (or any other) limit point,

$\text{dist}(x,M) \leq \|x-m\| \leq \limsup \|x-m_n\| = \lim \|x-m_n\| = \text{dist}(x,M)$, therefore $m \in \mathcal{P}_M(x)$.

The standard **example** for M is a finite dimensional linear subspace of X , e.g.,

$$\pi_n := \text{polynomials of degree} \leq n$$

as a subspace of $X = C[a,b]$ or $L_2[a,b]$. The desired local compactness is obvious for such an M . If M is a **non**linear subset, e.g., $M = \pi_n/\pi_m$:= rational functions of degree n over m , the argument becomes more sophisticated: The convergence notion used is weakened sufficiently to gain local compactness while not losing the semicontinuity of the norm with respect to this notion of convergence.

A real difficulty in multivariate approximation is the fact that it becomes reasonable to consider **infinite** dimensional M . E.g.,

$$M = C[a,b] + C[c,d] \subseteq C([a,b]\times[c,d])$$

provides a simple example of the reasonable attempt to approximate a function of many variables (in this case, two) by composition of functions with fewer variables (in this case, the sum of two functions of one variable). Now even existence is a nontrivial matter.

Uniqueness and **characterization** involve a ball game, of sorts. Imagine the **closed ball** $\overline{B}_r(x)$ **of radius** r **around** x . Starting with $r = 0$, let r grow until $r = r^* := \text{dist}(x,M)$. Then

$$\mathcal{P}_M(x) = M \ \overline{B}_{r^*}(x) .$$

For general M and some x , this first touch may well happen at two or more places. In such a circumstance, **local** uniqueness and characterization of a **local(ly)** b.a. become interesting questions. The interesting exception to this general statement is provided by a **convex** M , in which case we have the following picture:

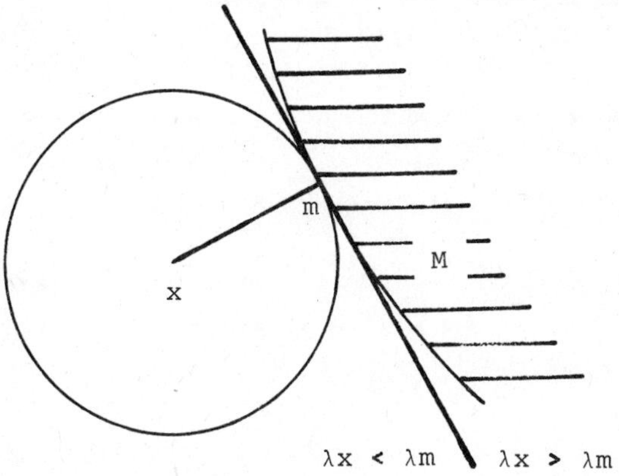

Geometric fact. *If* M *is convex and* $m \in M$ *and* $r := \|x-m\|$, *then* $m \in \mathcal{P}_M(x)$ *iff* *it is possible to separate* M *and* $B_r(x)$ *by a closed hyperplane through* m , *i.e., iff there exists* $\lambda \in X^*$ *for which*

$$\lambda[B_r(x)] < \lambda m \leq \lambda[M] \tag{1.1}$$

The first inequality in (1.1) is equivalent to

$$\lambda[B_r(0)] \quad < \quad \lambda(m-x) \ ,$$

i.e., to

$$\lambda \neq 0 \quad \text{and} \quad \|\lambda\| r \leq \lambda(m-x) \quad \left(\leq \|\lambda\| \|m-x\| = \|\lambda\| r \right) ,$$

hence to

$$\lambda \neq 0 \quad \text{and} \quad \lambda(m-x) = \|\lambda\| \|m-x\| \ .$$

This last condition is called variously ($\lambda \neq 0$ and) " $m-x$ *is an extremal for* λ " or " λ *takes on its norm on* $m-x$ " or " λ *is parallel to* $m-x$ ". I like this last phrase best and therefore write this condition

$$\lambda \ || \ m-x$$

to remind you of the familiar picture: In a finite dimensional setting, λ would simply be the vector **normal** to the (separating) hyperplane and would point in the same direction as $m-x$ does; in short, λ would be parallel to $m-x$.

As a matter of convenience, one talks instead about the **error** $x-m$. This requires switching the sign of λ and so gives the

Characterization Theorem. *Let* M *be a convex subset of the normed linear space* X , *let* $x \in X \setminus \overline{M}$, *and* $m \in M$. *Then*

(i) $m \in \mathcal{P}_M(x) \iff \exists \lambda \ || \ x-m$ *s.t.* $\lambda[M] \leq \lambda m$.

(ii) *If* M *is a linear subspace, then*

$m \in \mathcal{P}_M(x) \iff \exists \lambda \ || \ x-m$ *s.t.* $\lambda \perp M$ (i.e., $\lambda[M] = \{0\}$) .

Here is a good **exercise** which can be handled by entirely elementary means: Let $x \in X$, a nls, let $\lambda \in X^* \setminus 0$ and $m \in \ker \lambda$. Then, $m \in \mathcal{P}_{\ker \lambda}(x) \iff \lambda(x-m) = \|\lambda\| \|x-m\|$. For it, you might want to prove first that

$$x \in X \ , \ \lambda \in X^* \implies |\lambda x| = \|\lambda\| \operatorname{dist}(x, \ker \lambda) \tag{1.2}$$

which contains all the customary error estimates of elementary numerical analysis.

Since $\lambda \parallel x-m$ and $\lambda \perp M$ together imply that $\lambda \parallel x-m'$ for all $m' \in \mathcal{P}_M(x)$, **nonuniqueness** in case of a linear M is tied to the possibility of such a λ being parallel to more than one element (of the same size), i.e., for the hyperplane

$$\{y \in X : \lambda y = 1\}$$

to touch the unit ball $B_1(0)$ at more than one point. Since both M and $\overline{B}_{dist(x,M)}(x)$ are convex, having two distinct points m and m' in $\mathcal{P}_M(x)$ implies that the whole line segment $[m,m']$ between m and m' is in $\mathcal{P}_M(x)$. This says that the unit ball must contain line segments in its boundary, which is the same as saying that the norm is not **strictly convex.** Put positively, strict convexity of the norm (such as the $\mathbf{L}_p$-norm for $1 < p < \infty$) implies uniqueness of b.a. from a convex subset.

For more specific choices of X and M, the characterization theorem can be made more explicit, to the point where it can be used for the **construction** of a b.a. For example, if $X = C(T)$ and M is an n-dimensional linear subspace of X and $m \in M$, then

$m \in \mathcal{P}_M(x)$ <===>

$$\exists\, r \leq n, (w_i), (t_i) \text{ s.t. } \lambda := \Sigma_1^{r+1} w_i [t_i] \parallel x-m \quad \text{and} \quad \lambda \perp M .$$

Here, $[t]$ denotes the linear functional of point evaluation at t, i.e., $[t]:f \longmapsto f(t)$. Behind this specialization of the general characterization theorem is the result useful for Numerical Analysis that any linear functional on an (n+1)-dimensional subspace of $C(T)$ has a norm preserving extension of the form $\Sigma_1^{n+1} w_i [t_i]$.

You will recognize in this characterization the familiar statement that the error $x-m$ in a b.a. must take on its norm at points $t_1, \ldots, t_{r+1}$ with $r \leq n$ and such that, for some weights w_i with $w_i(x-m)(t_i) \geq 0$, all i, one has $\Sigma_1^{r+1} w_i [t_i] \perp M$.

It is not difficult to see that nonuniqueness is connected with having $r < n$ here. (Also, the norm is not strictly convex, so we would expect nonuniqueness for some x and M.) Recall that an n-dimensional subspace M of $C(T)$ is called a **Haar** space if, for any distinct points $t_1, \ldots, t_n$ in T, $([t_i])_1^n$ is linearly independent over M. For a Haar space M, having $0 \neq \lambda := \Sigma_1^{r+1} w_i [t_i] \perp M$ implies that $r \geq n$. Having in addition that $\lambda \parallel x-m$ implies that $\lambda \parallel x-m'$ for any other b.a. m', therefore m and m' must agree at the points $t_1, \ldots, t_{r+1}$ (assuming without loss that $w_i \neq 0$, all

i) , and, using once more that M is Haar, this implies that $m = m'$. Conversely, one can show that, if M is not Haar, then there are functions with many b.a.'s from M .

This equivalence between uniqueness and the Haar property has unhappy consequences for multivariate approximation, because of the following

Fact (Mairhuber). If T is not essentially just an interval, and if $\dim M > 1$, then M is not Haar.

The **proof** consists of a bit of railroading: Let $(f_1, \ldots, f_n)$ be a basis for M . Then $\det\big(f_j(t_i)\big)$ is a continuous function of the n points $t_1, \ldots, t_n$. If now T contains a "Y" , i.e., a "fork" or "switch", then one can continuously deform $(t_1, t_2, t_3, \ldots, t_n)$ into $(t_2, t_1, t_3, \ldots, t_n)$ while keeping the t_i's distinct:

$$\begin{array}{c} t_1 \\ t_2 \\ \cdot \\ t_n \end{array} \longrightarrow \begin{array}{c} t_1 \\ t_2 \\ t_3 \\ \cdot \\ t_n \end{array} \longrightarrow \begin{array}{ccc} t_1 & & t_2 \\ & t_3 & \\ & \cdot & \\ & t_n & \end{array} \longrightarrow \begin{array}{c} t_2 \\ t_1 \\ t_3 \\ \cdot \\ t_n \end{array} \longrightarrow \begin{array}{c} t_2 \\ t_1 \\ \cdot \\ t_n \end{array}$$

This means that the determinant has changed sign along the way, hence must have vanished for some choice of n distinct points.

The resulting nonuniqueness of b.a. in $C(T)$ for multidimensional T has produced a great industry in uniform approximation by functions of several variables and much fun can be had. I shall resist the temptation to enter into details now, because I am not convinced that **best** approximation is all that practical.

The question of **a priori bounds** or **degree of approximation** is concerned with

$$\operatorname{dist}(K,M) := \sup_{x \in K} \operatorname{dist}(x,M)$$

A typical example would be : $X = C[0,1]$, $M = \pi_n$, and $K := \{ x \in X : \|x''\|_\infty \le 1 \}$. Actually, it is not easy or even useful to be precise without getting simply the number $\operatorname{dist}(K,M)$. The question of degree of approximation comes into its own when one has given a **scale** (M_h) or (M_n) of subsets with $h \longrightarrow 0$ or $n \longrightarrow \infty$ and then considers

$$E_K(h) := \operatorname{dist}(K, M_h)$$

as a function of h . One proves **direct** or **Jackson**-type theorems:

$$x \in K \implies \operatorname{dist}(x, M_h) \sim h^r$$

and tries to demonstrate their sharpness, if possible, by proving **inverse** or **Bernstein**-type theorems:

$$\mathrm{dist}(x, M_h) \sim h^r \implies x/\|x\| \in K .$$

Related is the question: Given that $x \in K$, is the scale (M_h) a good choice for approximating x ? What is one to judge by? Kolmogorov [1936] proposed some time ago that $\mathrm{dist}(x,M)/\dim M$ is a good measure. He introduced

$$d_n(K) := \inf_{\dim M \leq n} \mathrm{dist}(K,M) =: \text{the } \textbf{n-width of } K .$$

While it is not easy to find an **optimal** subspace, i.e., a subspace at which the infimum of the definition is taken on, one would at least like an **asymptotically optimal** scale (M_n), i.e., a scale with $\dim M_n = n$ for which $\mathrm{dist}(K,M_n) = O(d_n(K))$.

Once **effort** enters considerations of approximation (here in the form of the dimension of M, i.e., the degrees of freedom used in the approximation), one can, of course, ask more: Is it really worthwhile to construct **best** approximations, particularly when a near-best approximation is cheaply available? Here we call the linear map $A:X \longrightarrow M$ a **near-best** approximation scheme if

$$\exists \text{ const } \forall x \in X \quad \|x-Ax\| \leq \text{const } \mathrm{dist}(x,M) .$$

Here, $\text{const} = 1$ would be best possible. Ax would then be a b.a. for every x. This does happen in inner product spaces but hardly anywhere else except in very special circumstances. In any event, such A is necessarily a **linear projector** (with $\mathrm{ran}\, A = M$) since the inequality implies that $A_{|M} = 1$. These approximation schemes are the topic of the second lecture.

There are many books on Approximation Theory available. One of the most striking is Lorentz [1966] . Akhiezer [1967] summarizes the classical part. Both Cheney [1966] and Rivlin [1969] provide a careful modern introduction to the field while the two volumes of Rice [1964, 1969] bring quite a bit of additional material, especially on approximation from a nonlinear M. Powell [1981] and Schönhage [1971] each give a very interesting view of the subject.

2. Linear Interpolation

Linear projectors arise from interpolation, as I intend to make clear in this lecture. Fortunately for me, Ward Cheney is with us who has spent a good part of his professional life studying linear projectors. He will no doubt be ready to answer all questions left over after (or raised by) this lecture. Look for his publications (e.g., Morris & Cheney [1974]). See also the excellent book by Davis [1963].

The setup is quite simple: We have a linear space X of functions on some domain T and, correspondingly, the linear space X' of linear functionals on X. We have given $f_1, \ldots, f_m \in X$ and $\lambda_1, \ldots, \lambda_n \in X'$ and consider the

Task: Given $g \in X$, construct $Pg := \Sigma_1^m \alpha(j) f_j$ such that Pg **interpolates to** g **at** $\lambda_1, \ldots, \lambda_n$, i.e.,

$$\lambda_i Pg = \lambda_i g \ , \ i=1,\ldots,n.$$

For **example**, the specifics: $T = [a,b]$, $X = C[a,b]$, $f_j = ()^{j-1}$ and $\lambda_i = [t_i]$ describe the task of polynomial interpolation. Altering this to $\lambda_i : f \longmapsto \int_T f(x) f_i(x)\,dx$, all i, and choosing $m = n$ describes least-squares approximation by polynomials.

Our first observation is that this task does not depend on the individual functions $f_1, \ldots, f_m$ nor on the individual linear functionals $\lambda_1, \ldots, \lambda_n$, but only on their spans

$$F := \text{span}\,(f_i)_1^m := \{\Sigma_1^m \alpha(j)\, f_j : \alpha \in \mathbf{R}^m\}$$

and

$$\Lambda := \text{span}\,(\lambda_i)_1^n .$$

This is obvious for the f_i's since the very task is stated in terms of their span. As to Λ, observe that

$$\lambda_i g = \lambda_i h \ , \ i=1,\ldots,n \iff \forall \beta \in \mathbf{R}^n \ (\Sigma\beta(i)\lambda_i)g = (\Sigma\beta(i)\lambda_i)h .$$

We use the abbreviation

$$\text{LIP}(F,\Lambda)$$

for the **Linear Interpolation Problem given by** F **and** Λ, i.e., for the

Task: Given $g \in X$, find $Pg \in F$ s.t. $g = Pg$ on Λ.

Here, F and Λ are understood to be linear subspaces (finite dimensional) of X and X', respectively. We call $\text{LIP}(F,\Lambda)$ **correct** if the task has exactly one solution for every $g \in X$.

Now, having just gotten rid of the f_i's and λ_i's , it is convenient to reintroduce them, in a possibly refined form: Let $(f_i)_1^m$ be a **basis** for F and let $(\lambda_i)_1^n$ be a **basis** for Λ . Then

$\Sigma_j \alpha(j) f_j$ solves LIP(F,Λ) for given g <===>

$$\alpha \in \mathbb{R}^m \text{ solves } \Sigma_j (\lambda_i f_j)\alpha(j) = \lambda_i g \ , \ i=1,\ldots,n.$$

We conclude

Lemma. (i) LIP(F,Λ) <u>is correct</u> <===> <u>the</u> **Gramian** G $:= (\lambda_i f_j)_{i=1,j=1}^{m \ \ n}$ <u>is invertible</u>.

(ii) LIP(F,Λ) <u>is correct</u> ===> $Pg = \Sigma_j \alpha(j) f_j$ <u>with</u> $\alpha = A^{-1}(\lambda_i g)$.

The **proof** is linear algebra: Since (f_i) is linearly independent, uniqueness is equivalent to having A 1-1. Since (λ_i) is linearly independent, existence is equivalent to having A onto. Note that correctness implies $m = n$.

The map P defined by such a correct LIP is **linear** (as a composition of the linear maps $g \longmapsto (\lambda_i g) \longmapsto \alpha = G^{-1}(\lambda_i g) \longmapsto \Sigma_j \alpha(j) f_j$). Also, by uniqueness, $P_{|F} = 1$, hence $P^2 = P$, showing that P is a linear **projector.** Its range is

$$\text{ran } P \ = \ F \ = \ \{x \in X : Px = x\}$$

while its kernel or nullspace is

$$\ker P \ = \ \{x \in X : \lambda x = 0, \text{ all } \lambda \in \Lambda\} \ =: \ \Lambda_\perp \ = \ \text{ran}(1-P) \ .$$

The customary view of a linear projector is that it provides a **direct sum decomposition:**

$$x \ = \ Px \ + \ (1-P)x \ .$$

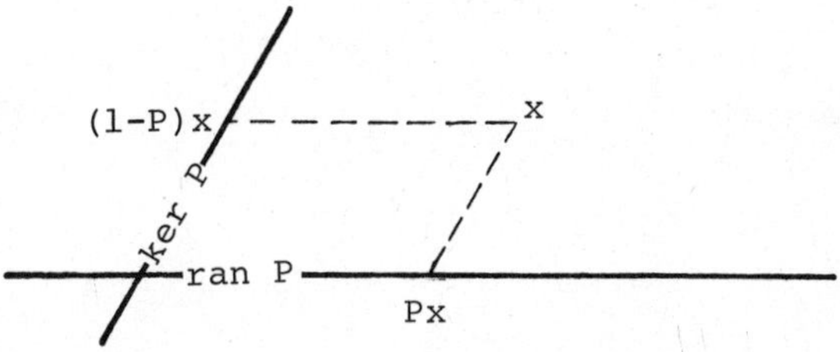

Px is the projection of x onto ran P along ker P . I prefer to think of P as **given by** F = ran P **and its interpolation conditions**

$$\Lambda \ = \ \{\mu \in X' : \mu P = \mu\} = \text{ran } P' \ .$$

This stresses the fact that Px is the unique element in ran P which agrees with x on Λ .

The **construction** of Pg involves, off hand, the solution of the linear system $G\alpha = (\lambda_i g)$. This can be viewed as switching over to the new basis (λ_i') for Λ with

$$\lambda_i' := \Sigma_j G^{-1}(i,j)\lambda_j .$$

Such a basis is, by its construction, **dual** to (f_i), i.e.,

$$\lambda_i' f_j = \delta_{ij} .$$

Another possibility is the **Lagrange** approach: Switch over to

$$f_j' := \Sigma_i G^{-1}(i,j) f_i$$

so that now $(\lambda_i f_j') = 1$. Yet another possibility is the **Newton** approach: If possible, switch over to

$$f_j' := \Sigma_i R^{-1}(i,j) f_i \quad , \quad \lambda_i' := \Sigma_j L^{-1}(i,j)\lambda_j$$

with LR a **triangular factorization** for G, giving again $(\lambda_i' f_j') = 1$. In this last approach, we would need $(\lambda_i f_j)_1^r$ invertible for $r=1,2,\ldots$. Equivalently, we would need the LIP(F_r,Λ_r) with $F_r :=$ span $(f_1,\ldots,f_r)$, $\Lambda_r :=$ span $(\lambda_1,\ldots,\lambda_r)$ to be correct, giving rise to a projector P_r, $r=1,2,\ldots,n$. In these terms, suitable bases for F and Λ can be constructed bootstrap fashion:

$$f_i' = (1-P_{i-1})f_i \quad , \qquad \lambda_i' = \lambda_i(1-P_{i-1})$$

$$P_i = \sum_{j<i} f_i'\lambda_i' / \lambda_i' f_i'$$

which is, in effect, Gauss elimination without pivoting.

The **example** inspiring all this terminology is, of course, **polynomial interpolation**, mentioned earlier, in which $f_j = (\,)^{j-1}$ and $\lambda_i = [t_i]$. The Lagrange approach leads to

$$f_j' : t \longmapsto \Pi_{i \neq j}(t-t_j)/(t_i-t_j)$$

while the Newton approach leads to

$$f_j' : t \longmapsto \Pi_{i<j}(t-t_j) \quad , \quad \lambda_i' = [t_1,\ldots,t_i] :=$$

the **divided difference at $t_1, \ldots, t_i$**, and thence to the Newton form

$$Pg = \Sigma_1^n [t_1,\ldots,t_i]g \prod_{j<i} (\cdot - t_j) .$$

Another well known example is specified by : $X = C(T)$, $F = \text{span}\ (f_i)$, of dimension n, and

$$\lambda_i : g \longmapsto \textstyle\int_T f_i g \quad , \quad i=1,\ldots,n .$$

Now P is **Least-squares** approximation, and the Newton approach is, in this instance, called **Gram-Schmidt orthogonalization.**

We are interested in linear projectors because they provide near-best linear approximation schemes. Explicitly, we have

Lebesgue's Inequality: $\|g - Pg\| \leq \|1-P\|\ \text{dist}(g,F)$

in case X is a normed linear space (=: nls) and P is bounded. In fact, we have a bit more:

$$|\mu g - \mu Pg| \le \operatorname{dist}(\mu, \Lambda)\ \|1-P\|\ \operatorname{dist}(g, F)\ ,\quad \mu \in X^*,\ g \in X\ . \tag{2.1}$$

For the **proof,** note that, for $\lambda \in \Lambda$ and $f \in F$, $\lambda(1-P) = 0$, $(1-P)f = 0$, so

$$|\mu g-\mu Pg| = |(\mu-\lambda)(1-P)(g-f)| \le \|\mu-\lambda\|\|1-P\|\|g-f\|\ .$$

Now take the infimum over $\lambda \in \Lambda$ and $f \in F$.

The inequality (2.1) is important for the **rule** makers who customarily approximate μg by μPg ever since Newton proposed this for $\mu g = \int g$ and P polynomial interpolation. The variational approach to splines, particularly important for the understanding of Duchon's multivariate 'thin plate' splines (Duchon [1976], [1977], Meinguet [1979]), takes off from this setup. See the epilogue.

The basic inequality (2.1) raises the two questions:
(i) Is P bounded? (ii) How big is $\|P\|$ (or, $\|1-P\|$)?
The following two lemmas give answers of sorts.

Lemma 2.1. Let P be given on the nls X by LIP(F,Λ). Then P is bounded iff $\Lambda \subseteq X^*$ (:= continuous linear functionals on X).

Proof. "==>" For all $\lambda \in \Lambda$, $\lambda = \lambda P = (\lambda_{|F})\circ P$ with $\lambda_{|F}$ continuous since $\dim F < \infty$.

"<==" We can write $P = \Sigma_1^n f_i' \lambda_i'$, hence $\|P\| \le \Sigma_1^n \|f_i'\|\|\lambda_i'\| < \infty$.
|||

Lemma 2.2. $\|P\| = \sup\limits_{f\in F}\ \inf\limits_{\lambda\in\Lambda}\ \|\lambda\|\|f\|/|\lambda f|$.

Proof. For any linear map A,

$$\|A\| = \sup_{x\notin \ker A} \|Ax\|/\|x\| = \sup_{x\notin\ker A}\ \sup_{y\in\ker A} \|Ax\|/\|x-y\|$$

$$= \sup_{x\notin\ker A} \|Ax\|/\operatorname{dist}(x,\ker A) = \sup_{x\notin\ker A}\ \inf_{\lambda\perp\ker A} \|Ax\|\|\lambda\|/|\lambda x|\ .$$

If now A is the linear projector P, then $x = Px + (1-P)x$ and $Px \in F = \operatorname{ran} P$ while $(1-P)x \in \ker P$. Further, $\lambda \perp \ker P$ iff $\lambda \in \Lambda$ (= ran P'), using the fact that $\operatorname{ran} P = \{x : x = Px\}$, hence $\lambda \perp \ker P \implies \lambda \perp \operatorname{ran}(1-P) \implies \lambda = \lambda P \implies \lambda \in \operatorname{ran} P' = \Lambda$.
Therefore,

$$\|P\| = \sup_{x\notin\ker P}\ \inf_{\lambda\perp\ker P} \|Px\|\|\lambda\|/|\lambda x| = \sup_{x\in F}\ \inf_{\lambda\in\Lambda} \|x\|\|\lambda\|/|\lambda x|\ . \quad |||$$

3. The tensor product construct

Our first foray into a multivariate setup is by tensor products. This construct is of limited use. Yet when it can be employed, it is so efficient that it is worth some effort to bring a given approximation problem into this form, if it can be done at all. Somewhat surprising approximation theoretic advantages of tensor products are discussed in de Boor & DeVore [1981].

The mathematics is quite simple, yet papers still appear which look forbidding and needlessly complicated since they do not make full use of the fact that, when dealing with tensor products, everything is essentially univariate, even the computer programs.

Naively, the tensor product of two univariate linear approximation schemes or projectors P and Q is formed as follows. For each fixed y, the linear projector P is applied to the **y-section**

$$h_y := h(\cdot,y)$$

of the bivariate function h, giving the univariate function

$$Ph_y = \Sigma_j \alpha(j;h_y)f_j$$

in which the coefficients $\alpha(j) = \alpha(j;h_y)$ depend on h_y, hence on y. Then Q is applied to each of these coefficient functions

$$c_j : y \longmapsto \alpha(j;h_y) ,$$

thus obtaining their univariate approximations

$$Qc_j = \Sigma_k \beta(k;j)g_k ,$$

with (g_k) a basis for $\operatorname{ran} Q$. Altogether, this gives the approximation

$$(Rh)(x,y) = \sum_{j,k} \beta(k;j)g_k(y)f_j(x)$$

to h.

Several questions are immediate: What is the nature of R? Is R a projector? What is the corresponding LIP? If we first use Q on each x-section of H and then P on the resulting coefficient functions, would the resulting approximation be again Rh?

We consider the last question first, since its answer supplies also the answer to the other questions. At its root is the question of whether the two operators $P\otimes 1$ and $1\otimes Q$ defined by

$$(P\otimes 1)h : (x,y) \longmapsto (Ph_y)(x) , \quad (1\otimes Q)h : (x,y) \longmapsto (Qh_x)(y)$$

commute. Since both P and Q are linear projectors, this question is settled once we know that $\lambda\otimes 1$ and $1\otimes\mu$ commute for each $\lambda \in \operatorname{ran} P'$ and each $\mu \in \operatorname{ran} Q'$. This is dealt with next in careful (and perhaps boring) detail.

Suppose that x_r is a ls of functions on some domain T_r, r=1,2. The tensor product $X_1 \otimes X_2$ of the two linear spaces X_1 and X_2 is customarily defined as the dual of the linear space of all bilinear functionals on $X_1 \times X_2$. Since I intend to use this concept only in the context of function spaces, I prefer the following definition: For $x_r \in X_r$,

$$x_1 \otimes x_2 : T_1 \times T_2 \longrightarrow \mathbf{R} : (t_1,t_2) \longmapsto x_1(t_1)x_2(t_2)$$

defines a function on $T_1 \times T_2$ called the **tensor product of** x_1 **and** x_2. Further,

$$X_1 \otimes X_2 \;:=\; \text{span}\{\, x_1 \otimes x_2 : x_r \in X_r\} \subseteq \mathbf{R}^{T_1 \times T_2} .$$

It is easy but essential to verify that the map

$$X_1 \times X_2 \longrightarrow X_1 \otimes X_2 : (x_1,x_2) \longmapsto x_1 \otimes x_2$$

is bilinear. This implies that

$$X_1 \otimes X_2 \;=\; \{\, \Sigma_j \, x_{1j} \otimes x_{2j} : x_{rj} \in X_r\}$$

(i.e., the scalars in the linear combinations making up $X_1 \otimes X_2$ can be absorbed).

Prime **examples** are the spaces

$\mathbf{P}_m \otimes \mathbf{P}_n$:= polynomials in two variables, of degree $< m$ in the first and of degree $< n$ in the second,

and

$$\mathbf{R}^m \otimes \mathbf{R}^n \;=\; \mathbf{R}^{m \times n} \;=\; \text{m-by-n matrices} .$$

Correspondingly, one defines the tensor product $\lambda \otimes \mu$ of $\lambda \in X_1'$ and $\mu \in X_2'$ as a linear functional on $X_1 \otimes X_2$ by the rule

$$\lambda \otimes \mu : X_1 \otimes X_2 \longrightarrow \mathbf{R} : \Sigma_j \, u_j \otimes v_j \longmapsto \Sigma_j (\lambda u_j)(\mu v_j) .$$

This is so obviously a linear functional on $X_1 \otimes X_2$ that it is easy to overlook the only nontrivial (though certainly not very deep) point about this definition, viz. whether it is a definition. The problem is that the rule for the value of $\lambda \otimes \mu$ at $w = \Sigma_j u_j \otimes v_j$ makes explicit use of the particular representation of w mentioned. Elements of $X_1 \otimes X_2$ have many different representations. For example, if $u = a+b$ and $v = 2c$, then

$$u \otimes v \;=\; a \otimes v \;+\; b \otimes v \;=\; a \otimes v \;+\; (2b) \otimes c \;=\; a \otimes c + a \otimes c + b \otimes v = \ldots$$

We should therefore have, correspondingly, that

$$\begin{aligned}(\lambda u)(\mu v) &= (\lambda a)(\mu v) + (\lambda b)(\mu v) = \\ &= (\lambda a)(\mu v) + (2\lambda b)(\mu c) = \\ &= (\lambda a)(\mu c) + (\lambda a)(\mu c) + (\lambda b)(\mu v) = \ldots\end{aligned}$$

We must therefore verify that <u>the number</u> $\Sigma_j (\lambda u_j)(\mu v_j)$ <u>depends only on</u> λ, μ, <u>and</u> $w := \Sigma_j u_j \otimes v_j$, <u>and not on the particular representation for</u> w <u>used</u>, i.e., not on the individual u_j's and v_j's. For this, let

$$v := \Sigma_j(\lambda u_j)v_j .$$

Then $v \in X_2$, and

$$v(t_2) = (\Sigma(\lambda u_j)v_j)(t_2) = \Sigma(\lambda u_j)v_j(t_2)$$

$$= \lambda(\Sigma u_j v_j(t_2)) = \lambda w(\cdot,t_2)$$

showing that v is the λ**-section** of w ,

$$v = w_\lambda .$$

In particular, v depends only on w and λ . On the other hand, $\Sigma_j(\lambda u_j)(\mu v_j) = \mu w_\lambda$.

Note that we get by symmetry that also $\Sigma_j(\lambda u_j)(\mu v_j) = \lambda w_\mu$, showing the hoped-for commutativity.

We are ready to define the tensor product of two LIPs (did the Tiny Tim craze ever catch on here?): Assume that $F_r \subseteq X_r \subseteq \mathbf{R}^{T_r}$ and $\Lambda_r \subseteq X_r'$ give rise to the correct LIP(F_r,Λ_r) with corresponding invertible Gramian

$$G_r := (\lambda_{ri} f_{rj})_{i,j=1}^{n_r} .$$

Pick some space W of functions on $T_1 \times T_2$ which contains $F := F_1 \otimes F_2$. Further, pick some $\nu_{ij} \in W'$ for which

$$\nu_{ij|F} = (\lambda_{1i} \otimes \lambda_{2j})_{|F} \text{ , all } i,j ,$$

and set $\Lambda := \text{span}(\nu_{ij})$.

Then: (i) $(\bar{f}_{ij}) := (f_{1i} \otimes f_{2j})$ is a basis for F .

(ii) The LIP(F,Λ) is correct.

(iii) The corresponding interpolant Pw to $w \in W$ can be computed as

$$Pw = \sum_{i,j} \Gamma_w(i,j)\, \bar{f}_{ij}$$

with

$$\Gamma_w = G_1^{-1} L_w (G_2^{-1})^T \quad \text{and} \quad L_w(i,j) := \nu_{ij} w \text{ , all } i,j . \tag{3.1}$$

For the **proof**, any $w \in F$ can be written $w = \Sigma_{ij} \Gamma(i,j)\bar{f}_{ij}$ for some suitable coefficient matrix Γ . From this, we compute that

$$L_w(r,s) := \nu_{rs} w = \sum_{i,j} \Gamma(i,j)(\lambda_{1r} f_{1i})(\lambda_{2s} f_{2j})$$

$$= \sum_{i,j} G_1(r,i)\, \Gamma(i,j)\, G_2(s,j) = G_1\, \Gamma\, G_2^T .$$

This shows that $\Gamma = G_1^{-1} L_w (G_2^T)^{-1}$, i.e., Γ depends only on w . This proves (i). Further, it shows that $(\lambda_{ij} w) = L \iff \Gamma = G_1^{-1} L (G_2^T)^{-1}$, thus proving (ii) and (iii).

A simple **example** is provided by the tensor product of cubic Hermite interpolation, i.e., cubic interpolation at the four points a, a, b, b, with osculatory parabolic interpolation, i.e., parabolic interpolation at the three points c, c, d . The above description leaves considerable freedom in the choice of the ν_{ij} . A natural choice would be

$$
\nu_{ij}w :=
\begin{array}{c|ccc}
i \backslash j & 1 & 2 & 3 \\
\hline
1 & w(a,c) & w_y(a,c) & w(a,d) \\
2 & w_x(a,c) & w_{xy}(a,c) & w_x(a,d) \\
3 & w(b,c) & w_y(b,c) & w(b,d) \\
4 & w_x(b,c) & w_{xy}(b,c) & w_x(b,d)
\end{array}
$$

and this would require the function space W to consist of functions defined (at least) on the rectangle $[a,b] \times [c,d]$ (assuming that a < b and c < d) and to have first derivatives and the first cross deri- vative (in a pointwise sense). Further, the desired commutativity would require that $w_{xy} = w_{yx}$ for all w ∈ W .

The computational advantage of this construct over other means of approximation is considerable. It generally takes $O(N^3/3)$ operations to solve the linear system for the coefficients of the solution of a LIP using N degrees of freedom. This number can be reduced considerably in a univariate setup (such as in spline approximation) through the use of special bases which make the resulting system banded. This is much harder in a multivariate context. In any event, if the LIP(F_r, Λ_r) in- volves n_r degrees of freedom, r=1,2 , then their tensor product in- volves $n_1 n_2$ degrees of freedom. Yet, using (3.1), one can compute the requisite $n_1 n_2$ coefficients in $O(n_1^3/3 + n_1^2 n_2 + n_1 n_2^2 + n_2^3/3)$ operations, since it only involves solving n_2 systems of order n_1 with the same coefficient matrix, and n_1 systems of order n_2 with the same coefficient matrix. Further savings are possible because this reduction of the computations to the univariate context provides the opportunity to make use of whatever savings are available in that con- text. If, for example, the univariate schemes give rise to banded syst- ems, then their tensor product can be constructed in $O(n_1 n_2)$ operat- ions.

This leads to a point made in de Boor [1979], that it is possible to form the "tensor product" of the computer programs for the solution of the univariate "factor"-problems. Typically, one has available pro- grams $INTER_r(B,M,A)$ which take the input vector $B = (\lambda_{ri} g)_1^{n_r}$ with $M = n_r$, and produce from it the desired coefficient vector $A = \alpha$ for the interpolant $P_r g = \Sigma \, \alpha(j) f_{rj}$ to g . Two changes are required to make such a program amenable to tensor product computations: One extends it to allow B to be a collection of N input vectors, i.e., to allow B to be an M-by-N matrix having these N data vectors as its columns; this requires that also N be supplied on input. Correspondingly, the output A becomes a collection of N coefficient vectors, i.e., a matrix. But, for reasons that will be obvious in a moment, the resulting coefficient vectors should form the **rows** of that matrix, i.e., A should be an N-by-M matrix. This gives the extension $INTER_r'(B,M,N,A)$. With it, we can solve the tensor product of the two $LIP(F_r,\Lambda_r)$, $r=1,2$, by the two calls

```
CALL INTER1'(Lw,n1,n2,A)
CALL INTER2'(A,n2,n1,Γ)
```

The two programs could even be the same as, e.g., in bicubic spline interpolation. Further, once such extended programs have been written, it is possible to carry out tensor product interpolation using more than two factors. Finally, this formation of program "tensor products" is also helpful in the evaluation or other manipulation of the interpolant. In any event, the detailed programming effort takes place at the univariate level, just as in the mathematical analysis of the construct.

The **error** is easily obtained formally. Writing $E_r := 1 - P_r$ for the error in the linear approximation scheme P_r , we find

$$1 = (P_1+E_1)\otimes(P_2+E_2) = P_1\otimes P_2 + P_1\otimes E_2 + E_1\otimes P_2 + E_1\otimes E_2 . \quad (3.2)$$

This shows the error operator for the tensor product scheme $P_1\otimes P_2$ to be a sum of the univariate errors. The order of approximation is therefore no better than the worse of the two univariate schemes.

Now note that $P_rE_r = E_rP_r = 0$. This implies that any sum of terms from the right hand side of (3.2) gives a linear projector. In particular, Gordon [1969]$_{1,2}$ has proposed the use of the socalled **Boolean sum**

$$P_1\oplus P_2 := P_1\otimes 1 + 1\otimes P_2 - P_1\otimes P_2 = P_1\otimes P_2 + P_1\otimes E_2 + E_1\otimes P_2 .$$

The resulting approximation scheme is called **blending** since it uses interpolation conditions of the form $\lambda\otimes 1$ and $1\otimes\mu$, hence, in its simplest form, matches information along certain lines parallel to the axes and so constructs a surface by "blending" together certain curves. For blending, the error is the **product** of the univariate errors. This improvement over the tensor product is bought at a high price: An infinite amount of information about the function to be approximated is required. Gordon has dealt successfully with this problem by proposing that one first use a relatively dense but finite amount of information to construct good approximations to the required curves and then use these approximations in the final construct.

4. Multivariate polynomial interpolation

I begin with a review of the standard notation concerning polynomials in m variables. The notation is designed to make it all look just as in the univariate case. The general **polynomial of total degree < k** is , by definition, any linear combination

$$x \longmapsto \Sigma_{|\alpha|<k} A_\alpha x^\alpha$$

of the **monomials** $()^\alpha$ with $|\alpha| < k$. Here,

$$x^\alpha := x(1)^{\alpha(1)} \cdot \ldots \cdot x(m)^{\alpha(m)}$$

and the **length** $|\alpha|$ of the integer vector α is defined by

$$|\alpha| := \|\alpha\|_1 = \alpha(1) + \ldots + \alpha(m)$$

if, as we assume, all the components of α are nonnegative. For such an **index vector**, one sets

$$\alpha! := \alpha(1)! \cdot \ldots \cdot \alpha(m)!$$

and thereby recovers the binomial formula

$$(x + y)^\alpha = \sum_{\beta<\alpha} \binom{\alpha}{\beta} x^\alpha y^\beta \quad .$$

The partial **ordering** used here is componentwise:

$$\beta < \alpha := \text{for all } i,\ \beta(i) < \alpha(i) \ .$$

This gives **Leibniz' formula**

$$D^\alpha(fg) = \sum_{\beta<\alpha} \binom{\alpha}{\beta} (D^\beta f)(D^{\alpha-\beta} g)$$

for the derivative of a product. Here,

$$D^\alpha := D_1^{\alpha(1)} \ldots D_m^{\alpha(m)}$$

with D_i the partial derivative with respect to the ith argument. More generally, $p(D)$ is the constant coefficient differential operator

$$p(D) := \Sigma_\alpha A_\alpha D^\alpha$$

in case p is the polynomial

$$p = \Sigma_\alpha A_\alpha ()^\alpha \quad .$$

For the special linear polynomial

$$p{:}x \longmapsto x^*y := \Sigma_i x(i)y(i) \ ,$$

we write

$$D_y := \Sigma\, y(i) D_i$$

instead of D^*y for the resulting (unnormalized) derivative in the direction of y.

In one variable, it is convenient to talk about $\mathbf{P}_k$, the linear space of polynomials of **order** k, i.e., of degree $< k$, since its dimension is k and the optimal approximation order from $\mathbf{P}_k$ achievable on an interval of length h is h^k. In several variables, approximation order continues to be linked to (total) polynomial order, but the dimension and other interesting quantities are more easily expressed in terms of (total) degree rather than order. For this reason, I will concentrate on the linear space

$$\pi_k = \pi_k(\mathbf{R}^m)$$

of all polynomials of (total) degree $\leq k$ in m variables. It is not difficult to see that

$$\dim \pi_k(\mathbf{R}^m) = \binom{m+k}{m} .$$

Indeed, the rule

$$\alpha(r) := i(r) - i(r-1) - 1 , \; r=1,\ldots,m$$

sets up a 1-1 correspondence between

$$\{\alpha \in \mathbf{Z}_+^m : |\alpha| \leq k\}$$

and the set

$$\{I \subseteq \{1,\ldots,m+k\} : |I| = m\}$$

of cardinality $\binom{m+k}{m}$ if we let $i(1), \ldots, i(m)$ be the elements of I, in increasing order (and take $i(0) = 0$). Thus the generating 'sequence'

$$(()^\alpha)_{|\alpha|\leq k}$$

for π_k contains $\binom{m+k}{m}$ terms. On the other hand, this sequence is linearly independent since, e.g.,

$$[0]D^\beta()^\alpha = \alpha!\,\delta_{\beta\alpha} .$$

Note that

$$\dim(\pi_{k+1} \ominus \pi_k) = \binom{m-1+k}{m-1}$$

since $\{\alpha \in \mathbf{Z}_+^m : |\alpha| = k\}$ is in obvious 1-1 correspondence with $\{\alpha \in \mathbf{Z}_+^{m-1} : |\alpha| \leq k\}$. This reaffirms the well known identity

$$\binom{m+k}{m} = \Sigma_{r=0}^k \binom{m-1+r}{m-1} .$$

We now consider the $\mathrm{LIP}(\pi_k,T) := \mathrm{LIP}(\pi_k,\mathrm{span}([t])_{t\in T})$ with T a subset of $\mathbf{R}^m$. We call T **correct** (for π_k) if the $\mathrm{LIP}(\pi_k,T)$ is correct. Since π_k is not the tensor product of univariate polynomial spaces, it seems unlikely that we could employ the tensor product construct to obtain correct T's. Yet it is possible, as the following example, two-dimensional for simplicity, shows. Recall that the linear projector of polynomial interpolation at points $u_0, \ldots, u_k$ can be written in Newton form as

$$P_u = \Sigma_0^k \phi_{ui}[u_0,\ldots,u_i]$$

with

$$\phi_{ui}(x) := (x-u_0)\ldots(x-u_{i-1}) , \text{ all } i .$$

Therefore, any partial sum

$$R_I := \Sigma_{(i,j)\in I} \phi_{ui}\otimes\phi_{vj}[u_0,\ldots,u_i]\otimes[v_0,\ldots,v_j]$$

of its tensor product with P_v for some point sequence $v_0, \ldots, v_k$ is also a linear projector, given that $I \subseteq \{0,\ldots,k\}^2$. The range of R_I is somewhere in $\pi_k\otimes\pi_k$. To insure that it is actually $\pi_k(\mathbf{R}^2)$, choose $I = \{(i,j) : i+j \leq k\}$. With this choice, ran $R_I \subseteq \pi_k$ and equality must hold since, by just counting terms, we see that ran R_I has dimension $\binom{2+k}{k}$ which is dim $\pi_k(\mathbf{R}^2)$. It is now a nice **exercise** to verify that, for this choice of I , R_I solves the LIP(π_k,T) with $T := \{(u_i,v_j) \in \mathbf{R}^2 : i+j\leq k\}$.

The same construction works in m variables and so provides the only standard choice of correct point sets T for $\pi_k(\mathbf{R}^m)$. This is the **simplicial** choice which, up to a linear change of variables, is

$$T = \{(u_p,v_q,\ldots,w_r) \in \mathbf{R}^m : p+q+\ldots+r\leq k\}$$

with (u_p), (v_p), ..., (w_p) given sequences of real numbers. Note that an affine change of variables

$$x \longmapsto Ax + b$$

leaves π_k invariant, hence leaves invariant the collection of correct point sets T for π_k .

More general correct point sets can be generated with the aid of the following theorem due to Chung & Yao [1977], - and here I must thank Dr. A. Genz for pointing out this reference to me.

Theorem 4.1. <u>If the point set</u> $T \subseteq \mathbf{R}^m$ <u>has cardinality</u> dim $\pi_k(\mathbf{R}^m)$, <u>and, for every</u> $t \in T$, <u>there exist</u> k <u>distinct hyperplanes on which all points in</u> T <u>lie except for</u> t , <u>then</u> T <u>is correct for</u> $\pi_k(\mathbf{R}^m)$.

It is clear how one would prove this theorem: For each $t \in T$, we can find, by assumption, k m-vectors a_1 , ..., a_k and scalars b_1 , ..., b_k so that the k-th degree polynomial

$$L_t : x \longmapsto (a_1{*}x-b_1)\ldots(a_k{*}x-b_k)$$

vanishes on $T\setminus t$ but not at t . This implies that, for any given g , the function

$$\Sigma_{t\in T}\, g(t)\, L_t/L_t(t)$$

is a polynomial of degree $\le k$ which agrees with g on T. On the other hand, the 'sequence' $(L_t)_{t\in T}$ is linearly independent (since it is obviously independent over T), and, by assumption, contains exactly $|T| = \dim \pi_k$ terms, hence must be a basis for π_k and this establishes the uniqueness of the interpolating polynomial. In short, we have the generalization of Lagrange's way of treating univariate polynomial interpolation.

A particularly striking instance of such a correct set T are the "natural lattices" of Chung & Yao [1977] rediscovered recently by Dahmen & Micchelli [1980], and also by Hakopian [1981]$_2$. Pick n points $a_1, \ldots, a_n$ in $\mathbf{R}^m$ so that the points $0, a_1, \ldots, a_n$ are **in general position**, i.e., any $m+1$ of them are affinely independent. To recall, $m+1$ points $b_0, \ldots, b_m$ in $\mathbf{R}^m$ are **affinely independent** if their affine hull is all of $\mathbf{R}^m$, i.e., if $\mathrm{vol}_m \operatorname{conv} (b_i)_0^M \ne 0$. Then, for any subset I of $\{1,\ldots,n\}$ with $|I| = m$, there exists exactly one x_I for which

$$1 + a_i{*}x_I = 0 , \text{ all } i \in I .$$

(Indeed, since $0, (a_i)_{i\in I}$ are affinely independent, the sequence $(a_i)_{i\in I}$ must be linearly independent.) Further, for this x_I, we must have

$$1 + a_j{*}x_I \ne 0 , \text{ all } j \notin I$$

(since $1 + a_j{*}x_I = 0$ implies that $(a_i)_{i\in I}, a_j$ all lie in the hyperplane $\{x \in \mathbf{R}^m : 1+x\cdot x_I = 0\}$, hence are not affinely independent which, by assumption, is possible only if $j \in I$). We conclude that

$$T := \{x_I : I \subseteq \{1,\ldots,n\}, |I| = m\}$$

is correct for $\pi_k(\mathbf{R}^m)$ with $k := n-m$, since $|T| = \binom{n}{m} = \dim \pi_k$ and

$$L_I : x \longmapsto \prod_{j\notin I} \frac{1 + a_j{*}x}{1 + a_j{*}x_I} \in \pi_k$$

with $L_I(x_J) = \delta_{IJ}$.

It is a nice **exercise** to develop a Newton form for the resulting polynomial interpolation scheme. This leads to a particular generalization of divided differences quite different from the tensor product construction with which we began this discussion.

The Newton approach has recently been generalized by Gasca & Maeztu [1980]. Although the idea is proposed in $\mathbf{R}^m$, I shall follow its authors and discuss details only in $\mathbf{R}^2$. Start with a straight line $1 + a_1{*}x = 0$. (This is not quite the most general line, but that doesn't matter.) Add a bunch of lines $1 + a_{1i}{*}x = 0$, intersecting the first line at distinct points x_{1i}, $i=1,\ldots,m_1$. Form the polynomials

$$p_{1i} : x \longmapsto (1 + a_{11}*x) \dots (1 + a_{1,i-1}*x) \ , \ i=1,\dots,m_1 \ .$$

The $LIP(span(p_{1i}), span([x_{1i}]))$ is correct since the Gram matrix $(p_{1j}(x_{1i}))$ is triangular with nonzero diagonal, hence invertible.

Now add a second line $1 + a_2*x = 0$ intersecting the first line at a point other than the x_{1i}'s (if at all), and add a second bunch of lines $1 + a_{2i}*x = 0$, intersecting the second line at distinct points x_{2i}, $i=1,\dots,m_2$. Form the corresponding polynomials

$$p_{2i} : x \longmapsto (1 + a_1*x)(1 + a_{21}*x) \dots (1 + a_{2,i-1}*x) \ , \ i=1,\dots,m_2 \ .$$

Then (and this is the salient part of the construction), the single linear factor $(1 + a_2*x)$ in the p_{2i} makes them vanish at all the earlier interpolation points x_{1j}. With this, the matrix $(p_{rj}(x_{si}))$ in **lexicographic** order is triangular with nonvanishing diagonal, and thus the $LIP(span(p_{rj}), span([x_{si}]))$ is correct.

The general pattern is now clear. What is less clear is just what $span(p_{rj})$ might be and, in more than two variables, things become horrendous. Still, for certain regular choices (see Maeztu [1982]), $span(p_{rj})$ can be shown to coincide with π_k and the corresponding correct point set can be more general than the simplicial choice, but not more general than those covered by Theorem 4.1.

In his 1978 thesis (see Kergin [1978], [1980]), Paul Kergin proposes a totally different approach to multivariate polynomial interpolation which, in a way, gave impetus to all the material yet to be discussed in these lectures. I begin with Kergin's result as he stated it.

Theorem 4.2. For any point sequence $t_0, \dots, t_n$ in $\mathbb{R}^m$ there exists exactly one map $P : C^{(n)}(\mathbb{R}^m) \longrightarrow \pi_n$ so that

(i) P is linear;

(ii) $\forall g \in C^{(n)}$ $\forall 0 \le k \le n$ $\forall q_k \in \pi_k$ homogeneous of degree k $\forall J \subseteq \{0,\dots,n\}$ with $|J| = k+1$,

$$q_k(D)Pg = q_k(D)g$$

at some point in $conv(t_j)_{j \in J}$.

Here, as earlier, $conv\ T$ denotes the convex hull of the point set T. Further, $q_k \in \pi_k$ is **homogeneous of degree k** in case

$$q_k = \sum_{|\alpha|=k} A_\alpha(\,)^\alpha \ .$$

Consider for a moment the special case $m = 2$. The stated requirement for $k = 0$ forces Pg to agree with g at each of the t_i's. For $k = 1$, we have

$$q_k(D) = A_{(1,0)}D_1 g + A_{(0,1)}D_2 g \ .$$

Thus, if $t_i \neq t_j$, then Pg would have to match any such derivative $q_1(D)$ somewhere on the segment between t_i and t_j. This condition is already satisfied in case $q_1(D)$ is in the same direction as the segment since Pg matches g at t_i and t_j. Therefore, this imposes just one additional condition, viz. that the derivative normal to the segment be matched somewhere along the segment. If $t_i = t_j$ for $i \neq j$, we get <u>two</u> additional conditions, viz. that Pg have the same tangent plane at t_i as does g. In other words, we obtain osculatory interpolation.

An extreme case of osculatory interpolation occurs in case $t_0 = \ldots = t_n$. Now Pg is necessarily just the Taylor expansion of degree $\leq n$ for g at t_0.

Kergin begins the **proof** of his theorem with the observation that P is necessarily **continuous** on $X := C^{(n)}(G)$ for any bounded G containing $t_0, \ldots, t_n$. Since P is linear, this requires only to show that P is bounded (on X). This latter fact Kergin shows by observing that, by assumption, the leading coefficients of Pg agree with the corresponding normalized derivatives of g at certain points, hence can be bounded in terms of $\|g\|_X$. He then considers Pg - Lg, with Lg the leading terms of Pg just estimated, and observes that the leading terms of the resulting polynomial are of lower order and must interpolate to the corresponding normalized derivatives of g - Lg at certain points, hence can be bounded in terms of $\|g - Lg\|_X$, therefore in terms of $\|g\|_X$. The inductive argument is now clear.

In consequence, P can be understood entirely from its action on a **fundamental subset** of $C^{(n)}$, i.e., a subset R whose finite linear combinations are dense in $C^{(n)}$. Kergin chooses R to consist of socalled **plane waves** (F. John) or **ridge functions** (C. A. Micchelli),

$$R := \{g \circ \lambda : g \in C^{(n)}(\mathbf{R}), \lambda \in (\mathbf{R}^m)'\}.$$

Such a function is constant in all planes normal to a certain direction. Explicitly,

$$(g \circ \lambda)(x) = g(\lambda^* x) = g\left(\Sigma_1^m \lambda(i) x(i)\right).$$

Note that it is sufficient to take just one suitable g, e.g., $g : t \longmapsto e^{it}$.

Next, Kergin shows **uniqueness.** To be sure, the claim is not that, for a given g, there is a unique $Pg \in \pi_n$ satisfying the conditions described in (ii), for that is not true. For example, the function $g : x \longmapsto x(1)x(2)$ has all functions $p : x \longmapsto ax(2)$ with $0 \leq a \leq 1$ as linear "interpolants" at the points (0,0) and (1,0) in that sense. Rather, Kergin claims the uniqueness of such a linear map P and proves it by showing that plane waves (with the univariate function g a polynomial) have unique "interpolants". Given the many conditions P has to satisfy, the uniqueness is not surprising. The hard part is to show **existence.**

For this, Kergin introduces (in rather different notation) the linear functionals

$$\int_{[x_0,\ldots,x_k]} g \quad := \quad \int_0^1 \cdots \int_0^{s_k} g(x_0+s_1\nabla x_1+ \cdots + s_k\nabla x_k)\, ds_k \cdots ds_1 \qquad (4.1)$$

and sets

$$Q \quad := \quad \operatorname{span}\{\, g \longmapsto \int_{[t_J]} q_k(D)g \;:\; |J| = k+1\,,\ k=0,\ldots,n\}\ .$$

Here, $t_J := (t_j)_{j\in J}$, with $J \subseteq \{0,\ldots,n\}$ as before. Then, by some hard counting, Kergin shows that $\dim Q \leq \dim \pi_n$. Add to this the fact that

$$Q_\perp \cap \pi_n \quad = \quad \{0\} \qquad (4.2)$$

and you can conclude that the $LIP(\pi_n, Q)$ is correct. Now take for P the resulting projector. Then

$$\int_{[t_J]} q_k(D)(g - Pg) \quad = \quad 0 \quad ,$$

hence $q_k(D)(g - Pg) = 0$ at some point in $\operatorname{conv} t_J$. The claim (4.2) is established by an inductive argument: If $p \in \pi_n \cap Q_\perp$, then, for all $|\alpha| = n$, $\int_{[t_0,\ldots,t_n]} D^\alpha p = 0$, therefore $D^\alpha p = 0$, i.e., $p \in \pi_{n-1}$, etc.

Micchelli & Milman [1980] give a striking formulation of Kergin's interpolation scheme which shows it to be a "lifting" of the Newton form of the univariate interpolating polynomial. Micchelli came to this by noticing that Kergin's linear functionals (4.1) are closely related to the divided difference (as the notation used in (4.1) already intimates) via the **Hermite-Genocchi formula** (see Nörlund [1923; p.16]):

$$[\tau_0,\ldots,\tau_k]g \quad = \quad \int_{[\tau_0,\ldots,\tau_k]} D^k g$$

for any sufficiently smooth univariate g, a fact easily proved by induction. This allows us to write the univariate polynomial interpolant $P_\tau f$ in Newton form as

$$(P_\tau g)(x) \quad = \quad \Sigma_0^n\, (x-\tau_0) \cdots (x-\tau_{k-1}) \int_{[\tau_0,\ldots,\tau_k]} D^k g \quad .$$

Also, recall that $[\tau_J]P_\tau g \;=\; [\tau_J]g$ for all $J \subseteq \{0,\ldots,n\}$. Now consider the Micchelli-Milman definition

$$Pf : x \longmapsto \sum_{k=0}^{n} \int_{[t_0,\ldots,t_k]} D_{x-t_0} \cdots D_{x-t_{k-1}} f \tag{4.3}$$

for any $f \in X := C^{(n)}(\mathbb{R}^m)$. The resulting map P is linear and continuous on X , hence can be understood by looking at its action on the set R of plane waves. For $f = g\circ\lambda \in R$, one computes

$$D_y f = \Sigma_1^m\, y(i)\, g^{(1)}(\lambda^*\cdot)\lambda(i) = (\lambda^* y)\, g^{(1)}\circ\lambda ,$$

therefore

$$D_{x-t_0} \cdots D_{x-t_{k-1}} f = \lambda^*(x-t_0) \cdots \lambda^*(x-t_{k-1})\, g^{(k)}\circ\lambda$$

and so

$$Pf(x) = \Sigma_0^n\, \lambda^*(x-t_0) \cdots \lambda^*(x-t_{k-1}) \int_{[\lambda^* t_0,\ldots,\lambda^* t_k]} g^{(k)} .$$

The last integral equals $[\lambda^* t_0,\ldots,\lambda^* t_k]g$, by the Hermite-Genocchi formula. Therefore, finally,

$$P(g\circ\lambda) = (P_{(\lambda^* t_i)}g)\circ\lambda .$$

This is the crucial observation. It shows that ran $P \subseteq \pi_n$ and that, for any $f = g\circ\lambda \in R$, any polynomial q_k homogeneous of degree k and any $J \subseteq \{0,\ldots,n\}$ with $|J| = k+1$,

$$\int_{[t_J]} q_k(D)Pf = \int_{[t_J]} q_k(D)f ,$$

since, for such an f ,

$$q_k(D)f = \sum_{|\alpha|=k} A_\alpha D_1^{\alpha(1)} \cdots D_m^{\alpha(m)} f = \sum_{|\alpha|=k} A_\alpha \lambda^\alpha g^{(k)}\circ\lambda$$

and

$$\int_{[t_J]} (P_{(\lambda^* t_i)}g)^{(k)}\circ\lambda = [\lambda^* t_J]P_{(\lambda^* t_i)}g = [\lambda^* t_J]g .$$

This establishes that (4.3) is a formula for Kergin's map. |||

Micchelli [1980] offers additional detail, e.g., the error formula one associates with the Newton form which leads to a constructive proof of the Bramble-Hilbert lemma (see Bramble & Hilbert [1970]).

Kergin's scheme raises some questions. In contrast to its univariate antecedent, it requires derivative information even when all the interpolation points are distinct. This has been remedied recently in Hakopian $[1981]_1$ by the simple device of lowering the order of the derivatives appearing in the definition of the linear functionals which make up Kergin's interpolation conditions Q . One may also investigate other variants of Kergin's scheme. In particular, Kergin's scheme makes it very attractive to study the linear functionals (4.1), as a basis for a suitable definition of the divided differences of a function of several variables. This was, in fact, the consideration which led Kergin to his scheme in the first place.

The study of these linear functionals led C. A. Micchelli to the recurrence relations for multivariate B-splines and so opened up that fruitful area of research which is the topic of the remaining lectures.

5. Multivariate B-splines

Following the lead of Schoenberg [1965], the multivariate B-spline $M(\cdot|t_0,\dots,t_n)$ was defined in de Boor [1976] by the rule

$$M(x|t_0,\dots,t_n) \quad := \quad \frac{\mathrm{vol}_{n-m}(P^{-1}x)\cap[t_0,\dots,t_n]}{\mathrm{vol}_n\ [t_0,\dots,t_n]} \qquad , \quad x\in\mathbf{R}^m \quad ,$$

thus generalizing a particular characterization of the univariate B-spline due to Curry & Schoenberg [1966]. Here, $t_0, \dots, t_n$ are points in $\mathbf{R}^n$, $[K]$ is the convex hull of the point set K , and P is the canonical projector

$$P : \mathbf{R}^n \longrightarrow \mathbf{R}^m : x \longmapsto (x(i))_1^m \quad .$$

Further, $\mathrm{vol}_k(K)$ is the k-dimensional volume of the set K .

Such a B-spline is a nonnegative piecewise polynomial function of degree $\leq k := n - m$, its support is $[Pt_0,\dots,Pt_n]$, and it is in C^{n-m-1} as long as the **knots** $Pt_0 , \dots, Pt_n$ are in general position. All this will be proved shortly.

In 1978, C. A. Micchelli [1980] proposed an equivalent but more suitable and flexible definition of $M(\cdot|t_0,\dots,t_n)$ as the **distribution** on $C_0^\infty(\mathbf{R}^m)$ given by the rule

$$M(\cdot|t_0,\dots,t_n) : f \longmapsto \quad n! \int_{[t_0,\dots,t_n]} f\circ P \quad . \tag{5.1}$$

This definition makes sense even if the t_i's are not in general position. $M(\cdot|t_0,\dots,t_n)$ is a **function** (in $\mathbf{L}_\infty(\mathbf{R}^m)$) if and only if $\mathrm{vol}_m[Pt_0,\dots,Pt_n] \neq 0$ and, in that case,

$$\int_{\mathbf{R}^m} M(\cdot|t_0,\dots,t_n)f \quad = \quad n! \int_{[t_0,\dots,t_n]} f\circ P \quad . \tag{5.2}$$

More than that, Micchelli [1980] proved **recurrence relations** for these multivariate B-splines, of the following form.

Theorem 5.1. <u>(i) If</u> $x = \Sigma\, \alpha_i Pt_i$ <u>with</u> $\Sigma\alpha_i = 0$, <u>then</u>

$$D_x M(\cdot|t_0,\dots,t_n) \quad = \quad n\ \Sigma\ \alpha_i M(\cdot|t_0,\dots,t_{i-1},t_{i+1},\dots,t_n) \ .$$

<u>(ii) If</u> $x = \Sigma\, \alpha_i Pt_i$ <u>with</u> $\Sigma\alpha_i = 1$, <u>then</u>

$$(n-m)\ M(x|t_0,\dots,t_n) \quad = \quad n\ \Sigma\ \alpha_i M(x|t_0,\dots,t_{i-1},t_{i+1},\dots,t_n) \ .$$

These recurrence relations were proved almost simultaneously by a different approach by Dahmen [1979]$_1$ and have since then been given different proofs by Micchelli [1979], Höllig [1980], Hakopian [1980], de Boor & Höllig [1981] and perhaps others. I'll now give a version of this last proof, from de Boor & Höllig [1982], not only because, naturally, I like it best, but because it covers a more general situation than described so far.

To begin with, I have to clear up an inconsistency in the notation I have employed. This inconsistency shows up in (5.2) where both

$$\int_{\mathbb{R}^m} \qquad \text{and} \qquad \int_{[t_0,\dots,t_n]}$$

occur and $[t_0,\dots,t_n]$ is, off-hand, **not** meant as the convex hull of the points $t_0, \dots, t_n$, but rather as an indication that the integral is to be formed as described in (4.1). These two meanings only differ by a scalar factor, though, viz. the factor $n!\mathrm{vol}_n[t_0,\dots,t_n]$, hence could be made to coincide if, in (4.1), we multiplied the right hand side by $n!\mathrm{vol}_n [t_0,\dots,t_n]$. I settle this inconsistency instead by abandoning **from now on** the interpretation (4.1) and entirely rely on the more naive interpretation of

$$\int_{[t_0,\dots,t_n]}$$

as the integral over the convex hull of the points $t_0, \dots, t_n$.

Consider now, more generally, a polyhedral convex set B in $\mathbb{R}^n$, some linear map P into $\mathbb{R}^m$ and having B in its domain, and the distribution M_B on $\mathbb{R}^m$ defined by the rule

$$M_B : f \longmapsto \int_B f \circ P \quad , \tag{5.3}$$

i.e., as the **P-shadow** of B. The **simplex** spline $M(\cdot|t_0,\dots,t_n)$ results when $B = [t_0,\dots,t_n]/\mathrm{vol}_n[t_0,\dots,t_n]$ and P is the canonical map $\mathbb{R}^n \longrightarrow \mathbb{R}^m$, but it has already turned out to be fruitful to allow more general sets B and, usually as a simplification, more general linear maps P.

In any case, the recurrence relations can be proved in this generality. The relevant observation is that the boundary of such a convex polyhedral set consists again of convex polyhedral sets B_i, of one dimension lower, hence Stokes' Theorem can be used to relate $M := M_B$ to $M_i := M_{B_i}$. For this, we also need the corresponding outward normal ν_i to B at B_i relative to the affine hull of B, and an arbitrary point b_i in the affine hull of B_i. For simplicity, we assume that B is a **body**, i.e., B has $\mathbb{R}^n$ as its affine hull. With these assumptions and notations, the following theorem holds.

Theorem 5.2. (i) $D_{Pz}M = -\Sigma_i z^*\nu_i M_i$, all $z \in \mathbf{R}^n$.
(ii) $(n-m)M(Pz) = \Sigma_i (b_i - z)^*\nu_i M_i(Pz)$, all $z \in \mathbf{R}^n$.
(iii) $DM = (n-m)M - \Sigma_i b_i{}^*\nu_i M_i$.

The **proof** of (i) is immediate:

$$(D_{Pz}M)f = -\int_B (D_{Pz}f)\circ P = -\int_B D_z(f\circ P) = -\int_{\partial B} z^*\nu\, f\circ P$$
$$= -\Sigma_i \int_{B_i} z^*\nu_i\, f\circ P = -\Sigma_i z^*\nu_i M_i f .$$

This uses the fact that, by definition, the derivative D_yM of the distribution M is the distribution obtained by the rule $f \longmapsto M(-D_y f)$, and the standard interplay

$$D_y(f\circ P) = (D_{Py}f)\circ P$$

between differentiation and linear change of variables. This interplay also proves that

$$(Df)(Px) = (D_{Px}f)(Px) = (D_x(f\circ P))(x) = (D(f\circ P))(x) \qquad (5.4)$$

with D the first order differential operator given by the rule

$$Df := \sum_{j=1}^{r} F_i D_i f$$

in case f has its domain in $\mathbf{R}^r$, and with

$$(F_j f)(x) := x(j)f(x) \text{ , all } j .$$

Thus $(Df)(x) = (D_x f)(x)$, and the **adjoint** of D is $-\Sigma D_j F_j$. This is of use in proving (iii): We have

$$D_j F_j = 1 + F_j D_j ,$$

therefore

$$-(DM)f = \int_B (\sum_{j=1}^{m} D_j F_j f)\circ P = mMf + \int_B (Df)\circ P$$

and

$$\int_B \sum_{i=1}^{n} D_i F_i (f\circ P) = nMf + \int_B D(f\circ P) .$$

Here, the last integral in the first line equals the last integral in the second, by (5.4). Therefore,

$$(DM)f = (n-m)Mf - \sum_{i=1}^{n} \int_B D_i F_i (f\circ P) .$$

This settles (iii) since

$$\sum_{i=1}^{n} \int_B D_i F_i (f\circ P) = \sum_{i=1}^{n} \int_{\partial B} \nu(i) F_i (f\circ P) = \int_{\partial B} (\cdot^*\nu)\, f\circ P$$

and, on the facet B_i , the function $(\cdot^*\nu)$ is constant.

Finally, to prove (ii), conclude from (i) and (iii) that, for any z with $Pz = x$,

$$0 = (D - D_{Pz})M(x)$$
$$= (n-m)M(x) - \Sigma\, b_i{*}\nu_i\, M_i(x) + \Sigma\, z{*}\nu_i\, M_i(x) .$$

As an **exercise**, I specialize Theorem 5.2 to the situation of Theorem 5.1. This means that $B = [t_0,\ldots,t_n]$ and that we may set $B_i := [(t_j)\setminus t_i]$, $i=0,\ldots,n$. Then

$$M = \mathrm{vol}_n B\; M(\cdot|t_0,\ldots,t_n) ,$$

therefore

$$M_i = \mathrm{vol}_{n-1}B_i\; M(\cdot|t_0,\ldots,t_{i-1},t_{i+1},\ldots,t_n) .$$

Also,

$$(t_i-b_i){*}\nu_i\ \mathrm{vol}\ B_i = -\, n\ \mathrm{vol}\ B$$

showing that the coeffients which appear in (i)-(iii) are, for this case, closely related to the **barycentric** or **areal** coordinates associated with the simplex B. In any case, if $z = \Sigma_j\alpha_j t_j$ with $\Sigma_j\alpha_j = 0$, then, since ν_i is perpendicular to the affine hull of $(t_j)\setminus t_i$, we have $(t_j-b_i){*}\nu_i = 0$ for all $j \neq i$ and so

$$z{*}\nu_i = \Sigma_j\alpha_j(t_j-b_i){*}\nu_i = \alpha_i(t_i-b_i){*}\nu_i ,$$

therefore, from (i),

$$D_{Pz}M(\cdot|(t_j)) = D_{Pz}M/\mathrm{vol}\ B = \left(-\Sigma_i\ z{*}\nu_i\ M_i\right)/\mathrm{vol}\ B$$
$$= n\ \Sigma_i\ \alpha_i\ M_i/\mathrm{vol}\ B_i = n\ \Sigma_i\alpha_i M(\cdot|(t_j)\setminus t_i)$$

This proves (i) of Micchelli's Theorem 5.1, with $x = Pz$. For Theorem 5.1(ii), we have, with $x = Pz$, $z = \Sigma_j\alpha_j t_j$ and $\Sigma_j\alpha_j = 1$, that

$$(b_i-z){*}\nu_i = (b_i-\Sigma\alpha_j t_j){*}\nu_i = \Sigma\alpha_j(b_i-t_j){*}\nu_i = \alpha_i(b_i-t_i){*}\nu_i ,$$

therefore, from (ii),

$$(n-m)M(x|(t_j)) = \Sigma_i(b_i-z){*}\nu_i\ M_i(x)/\mathrm{vol}\ B$$
$$= \Sigma_i\ \alpha_i\ (b_i-t_i){*}\nu_i\ M_i(x)/\mathrm{vol}\ B = \Sigma_i\ \alpha_i M(x|(t_j)\setminus t_i) .$$

W. Dahmen has pointed out to me that, once one recognizes that the recurrence relations in Theorem 5.1 can be written as in Theorem 5.2 in terms of facet normals, then Theorem 5.2 can be derived from Theorem 5.1 using the fact that any convex polyhedral body can be **triangulated**, i.e., is the essentially disjoint union of simplices, any two of which are either disjoint or else have exactly a face in common.

The recurrence relation (iii) was first stated and proved (for simplices) by Hakopian [1980], in his simple proof of Theorem 5.1.

Particularly useful choices for B other than the simplex include cones and boxes. The choice of the cone

$$\{t_0 + \Sigma_1^n \alpha_i(t_i - t_0) : 0 \le \alpha_i , \text{ all } i\}$$

with vertex t_0 and generating rays $t_i - t_0$, $i=1,\dots,n$ leads to the **cone** spline $M = M_B$. This is the **truncated power** introduced and heavily used in Dahmen [1979], in direct generalization of the functions

$$\mathbf{R} \longrightarrow \mathbf{R} : x \longmapsto (x - t)_+^{k-1}$$

familiar from univariate spline analysis. Of course, every such cone spline can be obtained as a translate of the P-shadow of the standard cone

$$\mathbf{R}_+^n := \{z \in \mathbf{R}^n : 0 \le z(i) , \text{ all } i \}$$

for an appropriate choice of the linear map P.

Choice of the box or parallelepiped

$$\{t_0 + \Sigma_1^n \alpha_i(t_i - t_0) : 0 \le \alpha_i \le 1 , \text{ all } i\}$$

gives rise to the **box** spline introduced in de Boor & DeVore [1981] and further studied in de Boor & Höllig [1982]. Any such box spline is a translate of the P-shadow of the standard box

$$\{z \in \mathbf{R}^n : 0 \le z(i) \le 1 , \text{ all } i\}$$

for an appropriate choice of the linear map P.

Repeated application of the recurrence relation (i) provides the information that, for arbitrary vectors $y_1, \dots, y_r$,

$$D_{y_1} \cdots D_{y_r} M_B \in \text{span}\{ M_F : F \text{ is an } (n-r)\text{dim.face of } B\}.$$

Thus all r-th order derivatives of M_B are in $\mathbf{L}_\infty$ provided PF is m-dimensional for every (n-r)-dimensional face F of B. This allows the following conclusions, asserted earlier for the simplex spline:

(iv) $M_B \in \mathbf{L}_\infty^{(d)} \subseteq C^{(d-1)}$, with

$$d := \max \{ r : \dim PF = m , \forall\ (n-r)\text{-dim.faces } F \text{ of } B\} ,$$

and, for this d, $M_B \notin C^{(d)}$.

(v) If PB is m-dimensional, then M_B is a pp function of degree $\le k := n-m$, with $\text{supp}\, M_B \subseteq PB$. Indeed, any (n-k-1)-dimensional face F of B is mapped by P into some hyperplane in $\mathbf{R}^m$, hence all (k+1)st order derivatives of M_B have their support entirely on some hyperplanes. This implies that, on each connected component of the complement of

$$\{ PF : F \text{ is a face of } B , \dim PF < m \} ,$$

M_B agrees with some element of π_k.

6. Approximation from the span of multivariate B-splines

A solitary B-spline is of little use in approximation. Thus we consider now a whole collection $\mathbf{B} := (B)$ of polyhedral convex bodies in $\mathbf{R}^n$, each giving rise to its P-shadow M_B, and ask just how $\mathbf{B}$ should be chosen so that we get a "useful" family (M_B). We use the well known properties of Schoenberg's univariate B-spline (see, e.g., de Boor [1976]) as a guide.

A first useful property of univariate B-splines is that, properly normalized, they form a **partition of unity.** This is not so hard to achieve with multivariate B-splines. We take for P the canonical map $\mathbf{R}^n \longrightarrow \mathbf{R}^m$ and choose $\mathbf{B}$ as a **partition** of some **slab** $\mathbf{R}^m \times \Delta$ in $\mathbf{R}^n$, for some $\Delta \subseteq \mathbf{R}^k$ (with $k := n-m$, as before). Then

$$\sum_{\mathbf{B}} M_B(x) = \sum_{\mathbf{B}} \mathrm{vol}_k P^{-1}x \cap B = \mathrm{vol}_k P^{-1}x \cap \bigcup_{\mathbf{B}} B = \mathrm{vol}_k \Delta ,$$

i.e., we have a partition of the constant $\mathrm{vol}_k \Delta$. Choosing Δ to have $\mathrm{vol}_k \Delta = 1$ or else dividing each M_B by $\mathrm{vol}_k \Delta$, gives the desired partition of unity. We conclude at once that a continuous function f can be approximated from

$$S_{\mathbf{B}} := \mathrm{span}\, (M_B)_{B \in \mathbf{B}}$$

to within

$$\omega(f; |\mathbf{B}|) ,$$

with

$$|\mathbf{B}| := \sup_{B \in \mathbf{B}} \mathrm{diam}\, PB .$$

The simple approximation

$$\sum f(\tau_B)\, M_B$$

with $\tau_B \in B$, all B, is that accurate (**exercise**).

Can we do better in case f is smoother? From the univariate theory, we would expect to get

$$\mathrm{dist}(f, S_{\mathbf{B}}) = O(|\mathbf{B}|^{k+1}) \tag{6.1}$$

in case $f \in \mathbf{L}_\infty^{k+1}$. This we could conclude at once if we had available a **quasi-interpolant** Q for $S_{\mathbf{B}}$, i.e., a bounded linear map Q into $S_{\mathbf{B}}$ which is **local** and **reproduces** π_k. A typical specification of "local" would be to require that $Qf_{|C}$ depend only on $f_{|N(C)}$ with

$$N(C) := \bigcup \{ PB : PB \cap C \neq \emptyset \} .$$

Then, as in Lebesgue's inequality, we could conclude that

$$f - Qf = f-p - Q(f-p) , \text{ for all } p \in \pi_k ,$$

therefore

$$\|(f-Qf)_{|C}\| \leq \|(f-p)_{|C}\| + \|Q\| \|(f-p)_{|N(C)}\| ,$$

and so

$$\|(f - Qf)_{|C}\| \leq (1 + \|Q\|)\, \mathrm{dist}_{N(C)}(f,\pi_k) .$$

This leads to (6.1).

A first requirement for such an argument is that $\pi_k \subseteq S_{\mathbf{B}}$. This was established by Dahmen in $[1979]_2$ by a clever argument for the case that **B** is a **triangulation** of such a slab

$$S := \mathbf{R}^m \times \Delta ,$$

i.e., **B** consists of simplices. In effect, he deforms the slab appropriately to S' so that the function

$$p : x \longmapsto \mathrm{vol}_k\, P^{-1}x \cap S'$$

is (locally) a polynomial of degree k , yet S' is still triangulated by the simplices B' , with B' the simplex into whose vertices the deformation sent the vertices of B . The deformation only takes place in directions perpendicular to $\mathbf{R}^m$. Therefore P carries the vertices of B and the corresponding vertices of B' to the same points. Consequently,

$$M_B/\mathrm{vol}_n B = M_{B'}/\mathrm{vol}_n B' ,$$

and so

$$p = \Sigma_{B\in\mathbf{B}}(\mathrm{vol}\, B'/\mathrm{vol}\, B)\, M_B .$$

He is able to modulate the deformation sufficiently to obtain a basis for π_k in this way.

This description neatly avoids discussion of some very nontrivial details.

In $[1980]_2$, Dahmen uses $\Delta = [0,1]^k$. There, he uses his result to support the claim that (6.1) holds (called Theorem 3.1 there), even without benefit of a quasi-interpolant, but I cannot follow the argument (even though it is a generalization of a univariate argument I once made up). I must therefore go the way of quasi-interpolants. Here I come across a difficulty first observed by Dahmen $[1980]_2$: With the choice $\Delta = [0,1]^k$, the B-splines (M_B) are **linearly dependent.** Dahmen overcomes this difficulty through an averaging process which provides a quasi-interpolant bounded in terms of a local mesh ratio.

There is another, practical difficulty, though, with this setup. The B-splines M_B , their support, their sets of discontinuities, all depend on the triangulation **B** of the slab S . In practice, one would like to **start** with some triangulation T of $\mathbf{R}^m$ and then construct B-splines in

$$\pi^r_{k,T} := \pi_{k,T} \cap C^{(r)} ,$$

the space of all pp functions of degree $\leq k$ and smoothness r

associated with that triangulation. Further, one would like to find enough B-splines to staff a basis for $\pi^r_{k,T}$, just as one is able to do for $m = 1$ (which acounts for the "B" in their name).

This, as it turns out, is too much to ask for, for various reasons. At present, the nature of the spaces $\pi^r_{k,T}$ is not at all understood, except when $r < 1$. We don't even know exactly when we are entitled to a local basis for that space. We must therefore be satisfied merely to construct $M_B \in \pi^r_{k,T}$. But even that is not possible in general, since the sets of discontinuities of M_B are associated with the projections of faces of B , hence not arbitrarily choosable. In particular, even if B is a simplex, the connected components of the complement of the discontinuity set need not be simplices.

One can hope, though, to obtain, for a given triangulation T , a linearly independent collection (M_B) of B-splines of prescribed smoothness, whose subdivision is a **refinement** of T and whose span $S_{\mathbf{B}}$ contains π_k . It is not difficult to do this for the univariate B-splines. Take for Δ the standard simplex $[e_0,\dots,e_k]$ in $\mathbf{R}^k$ (with $e_0 := 0$, $e_j := (\delta_{ij})$ for $j > 0$) and triangulate $\mathbf{R}\times\Delta$ by the simplices

$$\tau_i \;:=\; [t_i \times e_{[i]}, \;\dots,\; t_{i+k}\times e_{[i+k]}] \;,\quad \text{all } i \;,$$

with (t_i) the given knot sequence in $\mathbf{R}$ and $[j]$ the remainder on division of j by $k+1$. It is worthwhile to visualize this construction for $k = 1$ and 2 . The construction uses the total ordering of $\mathbf{R}$ in an essential way, hence it is not easily generalized to the general case. Nevertheless, a construction of the desired type was found by Dahmen & Micchelli [1982] and independently by Höllig [1982]. The underlying idea is to get a B-spline basis (M_B) for $\pi_{k,T}$ and then to "pull apart their knots". Explicitly, with

$$\Delta \;=\; [e_0,\dots,e_k]$$

the standard simplex in $\mathbf{R}^k$ and

$$\tau \;:=\; [v_0,\dots,v_m]$$

any particular simplex in the triangulation T , the **simploid**

$$\tau\times\Delta$$

is triangulated using the **combinatorial product** $\tau\diamond\Delta$. I find it helpful to visualize this construction in the following way:

In the cartesian product $(v_i)\times(e_j)$, there are $\binom{m+k}{m}$ nondecreasing "paths" with endpoints (τ_0,e_0) and (τ_m,e_k) . A typical "path" is shown in the following figure.

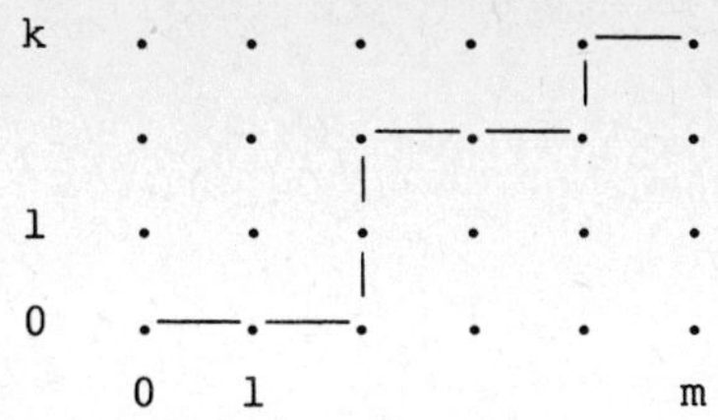

Associate with each such "path" the simplex σ in $\tau\times\Delta$ spanned by the points on that "path". It is a nice and worthwhile **exercise** to show that the resulting collection

$$\Sigma_\tau = (\sigma)$$

of simplices forms a triangulation for $\tau\times\Delta$. More than that, if we carry out this construction for each $\tau \in T$, making certain only that the ordering of the vertices of each τ used is **consistent**, i.e., comes from a total ordering of all the vertices appearing in T , then we obtain a triangulation

$$\Sigma \;:=\; T\diamond\Delta$$

for $\mathbb{R}^m\times\Delta$. This is straightforward except, perhaps, for the assertion that

$$\sigma\cap\sigma' \;=\; [V_\sigma\cap V_{\sigma'}] \quad , \quad \text{all } \sigma, \sigma' \in \Sigma ,$$

with V_σ the set of vertices of the simplex σ .

Now consider $(M_\sigma)_{\sigma\in\Sigma}$. For each $\sigma \in \Sigma_\tau$, M_σ has support in τ , and $PV_\sigma = V_\tau$, hence M_σ has τ as its support and agrees with some polynomial there. For a given $\tau \in T$, there are exactly $\binom{m+k}{m}$ such M_σ , i.e., exactly enough to staff a basis for $\pi_{k|\tau}$. Since, by Dahmen's result, $\pi_k \subseteq S_\Sigma$, it follows that these M_σ form a basis for $\pi_{k|\tau}$. We conclude that $(M_\sigma)_{\sigma\in\Sigma}$ <u>is a basis for</u> $\pi_{k,T}$.

It's time to pull apart the knots. For this, denote by

$$V(\Sigma) \;=\; V(T) \times \{e_0,\ldots,e_k\}$$

the vertices of the triangulation $\Sigma = T\diamond\Delta$. Let

$$F : V(\Sigma) \longrightarrow \mathbb{R}^n : (v,e_i) \longmapsto (\bar{F}v,e_i) \quad .$$

The only requirement Höllig [1982] makes on F (or $\bar{F}$) is that it be **locally finite**, i.e., $|\text{ran } F \cap C| < \infty$ for all bounded sets C . Dahmen & Micchelli [1982] require that $\text{sign det } \sigma = \text{sign det } F\sigma$, all $\sigma \in \Sigma$. Here,

$$\det \sigma$$

is the signed volume of the simplex σ . Its signature depends, of course, on the ordering of its vertices and the ordering is meant to be the ordering in $V(\Sigma)$.

With this, I am ready to state the basic result of this construction due to Dahmen & Micchelli and to Höllig.

Theorem 6.1. For all $y \in \mathbf{R}^m$,

$$(1 + x^*y)^k = \sum_{\sigma\in\Sigma} C_\sigma(y)\, M_{F\sigma}(x)$$

for all x at which all the simplex splines occuring on the right are continuous, with

$$C_\sigma(y) := k!\ \mathrm{sign}(\det \sigma)\ \det G_y F\sigma$$

and

$$G_y : \mathbf{R}^m \times \mathbf{R}^k \longrightarrow \mathbf{R}^n : (x,u) \longmapsto (x, (1+x^*y)u) \quad .$$

In effect, G_y carries out Dahmen's appropriate deformation of the slab $\mathbf{R}^m \times \Delta$ mentioned earlier.

Since y is arbitrary, we conclude that

$$\pi_k \subseteq S_{F\Sigma} \ .$$

The **quasi-interpolant** is now immediate, in case F is not too violent. Specifically, assume that, for each $\tau \in T$, there is some ball b_τ which is contained in the support of every $M_{F\sigma}$ with $\sigma \in \Sigma_\tau$ and is outside the support of every other $M_{F\sigma}$. This implies that

$$S_{F\Sigma} = \mathrm{span}\ (M_{F\sigma})_{s\in\Sigma_\tau} \quad \text{on} \quad b_\tau$$

and, as there are just enough simplex splines $M_{F\sigma}$ to staff a basis for π_k , this implies that

$$M_{F\sigma}\,|b_\tau \in \pi_k\,|b_\tau \quad \text{for every} \quad \sigma \in \Sigma_\tau \ .$$

This allows the construction of linear functionals λ_σ as normpreserving extensions, to $\mathbf{L}_1(b_\tau)$ say, of the coordinate functionals μ_σ on π_k which carry the polynomial $M_{F\sigma|b_\tau}$ to $\delta_{\sigma\sigma'}$, all $\sigma, \sigma' \in \Sigma_\tau$. Since

$$\mathrm{supp}\ \lambda_\sigma \subseteq b_\tau \ , \ \text{all} \quad \sigma \in \Sigma_\tau \ ,$$

this shows that (λ_σ) is dual to $(M_{F\sigma})$, i.e.,

$$\lambda_\sigma M_{F\sigma'} = \delta_{\sigma\sigma'} \ , \quad \text{all} \quad \sigma \in \Sigma \ .$$

The resulting quasi-interpolant

$$Q : f \longmapsto \Sigma\, \lambda_\sigma f\, M_{F\sigma}$$

is therefore even a **linear projector** onto $S_{F\Sigma}$, i.e., it reproduces all of $S_{F\Sigma}$ and not just π_k . The only concern is its size. $\|Q\|$ can be bounded in terms of the relative size of b_τ in τ . In particular, all is well in case the vertex perturbation map F is not too violent.

Theorem 6.1 provides a generalization of **Marsden's identity.** It is fair to say that it is based on a **two-dimensional** version of Marsden's identity first proved by Goodman & Lee [1981]. These authors provide, in particular, the more explicit formula

$$C_\sigma(y) = k! \prod_{j=3}^{n} (1 + z_j{*}y) \quad ,$$

with the z_j determined as follows: The vertices v_j of $F\sigma$ are of the form

$$Pv_j \times e_i$$

and, for each $i = 0,\ldots,k = n-2$, there is at least one j so that $v_j = (Pv_j, e_i)$. This leaves exactly two possibilities:

(i) for some i, there are three vertices of the form (Pv_j, e_i). Then we will call them v_0, v_1, v_2.

(ii) for two values of i, there are two vertices of the form (Pv_j, e_i). Then we will call one pair v_0, v_1 and the other v_2, v_3. With this, we take

$$z_j := \begin{cases} Pv_j & , j > 3 \\ Pv_3 \quad , \text{if (i)} & \\ \operatorname{aff}[Pv_0, Pv_1] \cap \operatorname{aff}[Pv_2, Pv_3] \quad , \text{if (ii)} & \Big\}, j = 3 \end{cases} \quad .$$

Höllig [1982] gives a simple example to show that such a nice formula with linearly factored coefficients is, in general, not to be expected for $m > 2$. Still, for the practically important case $m = 2$, these simple formulae lead Goodman & Lee to the intriguing generalization

$$V : f \longmapsto \Sigma f(t_\sigma) M_{F\sigma}$$

of Schoenberg's variation diminishing spline operator. This operator V is obviously positive regardless of the choice of the t_σ. Goodman & Lee choose

$$t_\sigma := (z_3 + \ldots + z_n)/(n-2)$$

since, in light of Theorem 6.1, this implies that V reproduces π_1. They are able to prove that, for any continuous f, Vf converges to f in the uniform norm as $|F\Sigma| \longrightarrow 0$, provided only that all the z_j's for σ lie in $PF\sigma$, an issue only for z_3 and only in case (ii).

There is an analogous quasi-interpolant construction for the span of certain translates of a **box spline** in de Boor & Höllig [1982]. The arguments have a different flavor, though, since the resulting pp functions have regular meshes, hence are amenable to "cardinal spline" techniques familiar from Schoenberg [1973].

7. Epilogue

In these lectures, I have touched on only very few questions of current interest in multivariate approximation theory. Even if I restrict attention to splines and pp functions, there are several areas of current research which I had intended to discuss when I first prepared for these lectures but which, in the end, I did not manage to fit into the allotted time.

The nature of the space

$$S := \pi^r_{k,T}$$

of smooth pp functions on a given partition T of some $\Omega \subseteq \mathbb{R}^m$ is not at all understood. Questions of interest concern the existence of a locally supported basis for S, the dimension of S, the dimension of the subspace consisting of those $f \in S$ which vanish $(r+1)$-fold at the boundary of Ω, the degree of approximation achievable from S. There is the **conjecture** that sufficiently smooth functions can be approximated to within $O(|T|^s)$ if and only if S contains a local partition of every $p \in \pi_{s-1}$, but attempts to prove this by construction of a quasi-interpolant have required, in addition, some kind of stability of the partitions. Work concerning $\dim S$ has been done only for $m = 2$ and initially only for $r = 1$, the first nontrivial case. See Strang [1974], Morgan & Scott [1975], and the survey of Schumaker [1979]. Most recently, Chui & Wang $[1981]_{1-3}$ have given precise results for certain T and arbitrary r. The existence of a local partition of unity in S is taken up in de Boor & DeVore [1981] for certain regular T in order to understand better the degree of approximation from S. These questions are further pursued in de Boor & Höllig [1982]. In both papers, the relationship between S and B-splines in S is explored, but even in the context of a simple and regular T (e.g., a rectangular grid with all north-east diagonals drawn in), this relationship is not yet fully understood.

The **adaptive** choice of the partition T is the topic of de Boor & Rice [1979]. Dahmen [1982] describes one way to use simplex splines adaptively. The degree of approximation achievable from S by proper choice of T is the topic of Dahmen, DeVore & Scherer [1980].

Practial aspects of approximation from $S = \pi^r_{k,T}$ on $\mathbf{R}^2$ are the topic of the two survey papers Barnhill [1977] and Schumaker [1976]. An interesting comparison of methods is given in Franke [1982]. And then there is the vast literature on the constructive aspects of the **Finite Element Method**! Some references of particular interest to Approximation Theory are : Ciarlet & Raviart $[1972]_{1,2}$, Courant [1943], Fix & Strang [1969], Guglielmo [1969] , Strang & Fix [1973], but this is clearly just a taste.

The **variational approach** to splines is, of course, not restricted to the univariate situation. Already Golomb & Weinberger [1959] consider particular bivariate examples as illustrations of the general theory. This theory has the following setting. A collection Λ of continuous linear functionals on some linear space X is given. Since the problem will only involve

$$\Lambda_\perp \; := \; \{x \in X : \lambda x = 0 \quad \text{for all} \quad \lambda \in \Lambda\} \quad ,$$

we might as well assume that Λ is a closed subspace of X^*. Further, a bounded linear map T from X to some normed linear space Y is given. The problem is to determine, for given $x \in X$, if possible, an element x^* at which the map

$$x + \Lambda_\perp \longrightarrow R_+ : y \longmapsto \|Ty\|$$

takes on its minimum. Such a minimizer x^* is called a **(T,Λ)-spline interpolant** to x . The word "interpolant" is appropriate since x^* agrees with x on Λ . Golomb & Weinberger [1959] deal with the special case: $X = Y$ is a Hilbert space and $T = 1$. In this setting, the map $x \longmapsto x^*$ is just the orthogonal projector onto Λ considered as a subspace of X . This schizophrenic nature of the interpolation conditions Λ , linear functionals on the one hand and elements of X on the other, is at the heart of the practical application of this theory. In standard Hilbert spaces of smooth functions on some domain Ω , the linear functional $[t]$ of evaluation at some point t turns into the section $G(t,\cdot)$ of the appropriate Green's function. In particular, when $X = L_2^{(k)}[a,b]$, then $[t]$ is represented by a pp function of degree $2k-1$ and in $C^{(2k-2)}$ with just one breakpoint, at t .

I was held back from exploring multivariate splines obtained in this way by the realization that this would require me to obtain and work with the Green's function relevant to X . This would usually not be polynomial nor locally simple, and would depend essentially on the domain Ω . Duchon [1976], [1977] dealt with such objections by the very effective device of choosing all of $\mathbf{R}^m$ for Ω . The resulting **thin plate splines** have already found practical use. Their theory is described invitingly in Meinguet [1979].

References

N. I. Akhiezer [1967], Vorlesungen über Approximationstheorie, 2., verbesserte Auflage, Akademie-Verlag, Berlin; appeared also as Theory of Approximation, F. Ungar Publ., New York, 1956.

R. E. Barnhill [1977], Representation and approximation of surfaces, in Mathematical Software III, J. Rice ed., Academic Press, New York, 69-120.

C. de Boor [1976], Splines as linear combinations of B-splines, in Approximation Theory II, G.G. Lorentz, C.K. Chui & L.L. Schumaker eds., Academic Press, 1-47.

C. de Boor [1979], Efficient computer manipulation of tensor products, ACM Trans.Math.Software 5, 173-182.

C. de Boor & R. DeVore [1981], Approximation by smooth multivariate splines, Math.Research Center TSR #2319. Trans.Amer.Math.Soc., to appear.

C. de Boor & K. Höllig [1981], Recurrence relations for multivariate B-splines, Math.Research Center TSR #2215. Proc.Amer.Math.Soc., to appear.

C. de Boor & K. Höllig [1982], B-splines from parallelepipeds, Math.Research Center TSR #2320.

C. de Boor & J. R. Rice [1979], An adaptive algorithm for multivariate approximation giving optimal convergence rates, J.Approx. Theory 25, 337-359.

J. H. Bramble & S. R. Hilbert [1970], Estimation of linear functionals on Sobolev spaces with applications to Fourier transforms and spline interpolation, SIAM J.Numer.Anal. 7, 112-124.

J. H. Bramble & S. R. Hilbert [1971], Bounds for a class of linear functionals with applications to Hermite interpolation, Numer.Math. 16, 362-369.

E. W. Cheney [1966], Introduction to Approximation Theory, McGraw-Hill, New York.

C. K. Chui & R.-H. Wang $[1981]_1$, Multivariate B-splines on triangulated rectangles, CAT # 6, Center for Approximation Theory, Texas A&M University, College Station, TX.

C. K. Chui & R.-H. Wang $[1981]_2$, On a bivariate B-spline basis, CAT # 7.

C. K. Chui & R.-H. Wang $[1981]_3$, Multivariate spline spaces, CAT #9.

K. C. Chung & T. H. Yao [1977], On lattices admitting unique Lagrange interpolations, SIAM J.Numer.Anal. 14, 735-741.

P. G. Ciarlet & R. A. Raviart $[1972]_1$, General Lagrange and Hermite interpolation in R^k with applications to finite element methods, Arch.Rat.Mech.Anal. 46,177-199.

P. G. Ciarlet & R. A. Raviart $[1972]_2$, Interpolation theory over curved elements, with applications to finite element methods, Computer Methods in Appl.Mech.Eng. 1, 217-249.

R. Courant [1943], Variational methods for the solution of problems of equilibrium and vibrations, Bull.Amer.Math.Soc. 49, 1-23.

H. B. Curry & I. J. Schoenberg [1966], Pólya frequency functions IV. The fundamental spline functions and their limits, J.d'Anal.Math. 17, 71-107.

W. Dahmen $[1979]_1$, Multivariate B-splines - recurrence relations and linear combinations of truncated powers, in Multivariate Approximation Theory, W. Schempp & K. Zeller eds., Birkhäuser, Basel, 64-82.

W. Dahmen $[1979]_2$, Polynomials as linear combinations of multivariate B-splines, Math.Z. 169, 93-98.

W. Dahmen $[1980]_1$, On multivariate B-splines, SIAM J.Numer.Anal. 17, 179-191.

W. Dahmen $[1980]_2$, Approximation by linear combinations of multivariate B-splines, J.Approx.Theory, to appear.

W. Dahmen [1982], Adaptive approximation by multivariate smooth splines, J.Approx.Theory, to appear.

W. Dahmen, R. DeVore & K. Scherer [1980], Multi-dimensional spline approximation, SIAM J.Numer.Anal. 17, 380-402.

W. Dahmen & C. A. Micchelli [1980], On limits of multivariate B-splines, Math.Research Center TSR # 2114. J.d'Anal.Math. 39 (1981),256-278.

W. Dahmen & C. A. Micchelli [1982], On the linear independence of multivariate B-splines.I, Triangulation of simploids, SIAM J.Numer.Anal. xx, xxx-xxx.

P. Davis [1963], Interpolation and Approximation, Blaisdell, Waltham MA. Now available from Dover.

J. Duchon [1976], Interpolation des fonctions de deux variables suivant le principe de la flexion des plaques minces, R.A.I.R.O. Analyse numerique 10, 5-12.

J. Duchon [1977], Splines minimizing rotation-invariant seminorms in Sobolev spaces, in Constructive Theory of Functions in Several Variables, Oberwolfach 1976, W. Schempp & K. Zeller eds., Springer-Verlag, Heidelberg, 85-100.

G. Fix & G. Strang [1969], Fourier analysis of the finite element method in Ritz-Galerkin theory, Studies in Appl.Math. 48, 265-273.

R. Franke [1982], Testing methods for scattered data interpolation and some results, Math.Comp. 38, 181-200.

M. Gasca & J. I. Maeztu [1980], On Lagrange and Hermite interpolation in R^k, Numer.Math., to appear.

F. di Guglielmo [1969], Construction d'approximations des espaces de Sobolev sur des reseaux en simplexes, Calcolo 6, 279-331.

M. Golomb & H. F. Weinberger [1959], Optimal approximation and error bounds, in On Numerical Approximation, R. Langer ed., U. of Wisconsin Press, 117-190.

T. N. T. Goodman & S. L. Lee [1981], Spline approximation operators of Bernstein-Schoenberg type in one and two variables, J.Approx.Theory 33, 248-263.

W. J. Gordon $[1969]_1$, Spline-blended surface interpolation through curve networks, J.Math.Mech. 18, 931-952.

W. J. Gordon $[1969]_2$, Distributive lattices and approximation of multivariate functions, in Approximations with special emphasis on spline functions, I. J. Schoenberg ed., Academic Press, New York, 223-277.

H. Hakopian [1980], On multivariate B-splines, SIAM J.Numer.Anal. xx, xxx-xxx.

H. Hakopian $[1981]_1$, Les differences divisees de plusieurs variables et les interpolations multidimensionelles de types lagrangien et hermitien, C.R. Acad.Sc.Paris 292 , 453-456.

H. Hakopian $[1981]_2$, Multivariate spline functions, B-spline basis and polynomial interpolations, ms.

K. Höllig [1981], A remark on multivariate B-splines, J.Approx. Theory 33, 119-125.

K. Höllig [1982], Multivariate splines, SIAM J.Numer.Anal. xx, xxx-xxx.

P. Kergin [1978], Interpolation of C^k functions, Ph.D. Thesis, University of Toronto, Canada.

P. Kergin [1980], A natural interpolation of C^k functions, J.Approx.Theory 29, 278-293.

A. N. Kolmogorov [1936], Über die beste Annäherung von Funktionen einer gegebenen Funktionenklasse, Ann.Math. 37, 107-111.

G. G. Lorentz [1966], Approximation of Functions, Holt, Rinehart & Winston, New York.

J. I. Maeztu [1982], Divided differences associated with reversible systems in R^2, SIAM J.Numer.Anal., to appear.

J. Meinguet [1979], Multivariate interpolation at arbitrary points made simple, J.Appl.Math.Phys. (ZAMP) 30, 292-304.

C. A. Micchelli [1980], A constructive approach to Kergin interpolation in $\mathbb{R}^k$: multivariate B-splines and Lagrange interpolation, Rocky Mountains J.Math. 10, 485-497.

C. A. Micchelli [1979], On a numerically efficient method for computing multivariate B-splines, in Multivariate Approximation Theory, W. Schempp & K. Zeller eds., Birkhäuser, Basel, 211-248.

C. A. Micchelli & P. Milman [1980], A formula for Kergin interpolation in $\mathbb{R}^k$, J.Approx.Theory 29, 294-296.

J. Morgan & R. Scott [1975], The dimension of the space of C^1 piecewise polynomials, ms.

P. D. Morris & E. W. Cheney [1974], On the existence and characterization of minimal projectors, J.reine angew.Math. 270, 61-76.

N. E. Nörlund [1924], Vorlesungen über Differenzenrechnung, Springer Grundlehren Bd. 13, Berlin.

M. J. D. Powell [1981], Approximation theory and methods, Cambridge University Press.

J. R. Rice [1964, 1969], The Approximation of Functions. Vols. I., II. , Addison-Wesley, Reading MA.

T. Rivlin [1969], In Introduction to the Approximation of Functions, Blaisdell, Waltham MA.

I. J. Schoenberg [1965], letter to Philip J. Davis dated May 31, 1965.

I. J. Schoenberg [1973], Cardinal Spline Interpolation, SIAM, Philadelphia PA.

A. Schönhage [1971], Approximationstheorie, de Gruyter, Berlin.

L. L. Schumaker [1976], Fitting surfaces to scattered data, in Approximation Theory II, G.G. Lorentz, C. K. Chui & L. L. Schumaker eds., Academic Press, New York, 203-268.

L. L. Schumaker [1979], On the dimension of spaces of piecewise polynomials in two variables, in Multivariate Approximation Theory, W. Schempp & K. Zeller eds., Birkhäuser, Basel, 396-412.

G. Strang [1974], The dimension of piecewise polynomials, and one-sided approximation, Springer Verlag Lecture Notes 363, 144-152.

G. Strang & G. J. Fix [1973], An Analysis of the Finite Element Method, Prentice-Hall, Englewood Cliffs.

PRACTICAL SPLINE APPROXIMATION

M.G. Cox
National Physical Laboratory
Teddington, Middlesex, TW11 OLW, UK

Abstract

This two-part paper describes the use of polynomial spline functions for purposes of interpolation and approximation. The emphasis is on practical utility rather than detailed theory. Part I introduces polynomial splines, defines B-splines and treats the representation of splines in terms of B-splines. Part II deals with the statement and solution of spline interpolation and least squares spline approximation problems. It also discusses strategies for selecting particular solutions to spline approximation problems having nonunique solutions and techniques for automatic knot placement.

Scope

Polynomial spline functions (or simply polynomial splines or splines) have diverse application. They have been used to provide solutions to mathematical problems in interpolation, data and function approximation, ordinary and partial differential equations, and integral equations. Splines have also been employed in many scientific and engineering applications; ones with which I personally have been concerned include instrument calibration, sonar signal analysis, highway visualization, terrain following, computer aided design and manufacture, radioimmunoassay, telescope design and plant growth analysis.

This paper places particular emphasis upon the algorithmic aspects of spline interpolation and least squares spline approximation. Additionally, the related tasks of evaluation, differentiation and indefinite integration of spline interpolants and approximants are discussed. Importance is attached to the use of a representation for *splines of general order* that affords a good balance between numerical stability and efficiency.

Consideration is also given to the solution of the systems of linear algebraic equations that arise in the construction of spline interpolants and approximants. The matrices associated with these equations are banded in form, and the elimination and orthogonalization techniques described take full advantage of this structure.

Splines are represented here in terms of a *basis* (i.e. as a linear combination of certain basis splines, just as polynomials can be expressed as a linear combination of certain basis polynomials such as Chebyshev or Legendre polynomials). Such a representation is used in preference to the redundant one consisting of a set of polynomial pieces together with continuity conditions at the joins. The truncated power functions form one possible basis, but the distinct advantages of employing instead a basis consisting of certain linear combinations of the truncated power functions - the B-splines - will be demonstrated.

References to equations take the form (e) or, if reference is to another part, p(e), where e is the equation number and p the part number.

PART I: POLYNOMIAL SPLINES AND THE B-SPLINE REPRESENTATION

In order to provide a framework for the discussion of polynomial splines some elementary concepts are first reviewed.

Divided differences

The nth divided difference of a function f at the points $\lambda_{j-n}, \ldots, \lambda_j$ is denoted by $[\lambda_{j-n}, \ldots, \lambda_j]f$. Divided differences are defined and computed recursively:

$$[\lambda_j]f = f(\lambda_j),$$

$$[\lambda_{j-n}, \ldots, \lambda_j]f = \begin{cases} f^{(n)}(\lambda_j)/n! & (\text{if } \lambda_{j-n} = \ldots = \lambda_j), \\ ([\lambda_{j-n}, \ldots, \lambda_j \backslash \lambda_k]f - [\lambda_{j-n}, \ldots, \lambda_j \backslash \lambda_\ell]f)/(\lambda_\ell - \lambda_k), & \\ \qquad k, \ell \in \{j-n, \ldots, j\}, \ \lambda_k \neq \lambda_\ell \ (\text{otherwise}). & \end{cases} \quad (1)$$

(The notation $\lambda_{j-n}, \ldots, \lambda_j \backslash \lambda_k$ represents the set of values $\lambda_{j-n}, \ldots, \lambda_j$ *less* λ_k.)

Divided difference table

Given function values $f(\lambda_1), \ldots, f(\lambda_n)$, (1) may be used to construct column by column the difference table:

$$\begin{array}{ccccc}
f(\lambda_1) & & & & \\
 & [\lambda_1,\lambda_2]f & & & \\
f(\lambda_2) & & [\lambda_1,\lambda_2,\lambda_3]f & \cdot & \\
 & [\lambda_2,\lambda_3]f & \vdots & \cdot & \\
f(\lambda_3) & \vdots & \vdots & & [\lambda_1, \ldots, \lambda_n]f \\
\vdots & \vdots & [\lambda_{n-2},\lambda_{n-1},\lambda_n]f & \cdot & \\
\vdots & [\lambda_{n-1},\lambda_n]f & & & \\
f(\lambda_n) & & & &
\end{array}$$

Leibnitz formula

The following identity, valuable in proving certain results for B-splines, is attributed to Leibnitz (de Boor, 1978). For given functions f and g,

$$[\lambda_{j-n}, \ldots, \lambda_j](fg) = \sum_{k=j-n}^{j} ([\lambda_{j-n}, \ldots, \lambda_k]f)([\lambda_k, \ldots, \lambda_j]g). \tag{2}$$

The knot set and its extension

Consider a finite interval $[x_{min}, x_{max}]$ and its partition

$$x_{min} < \lambda_1 \leqslant \ldots \leqslant \lambda_N < x_{max}. \tag{3}$$

Example N=9

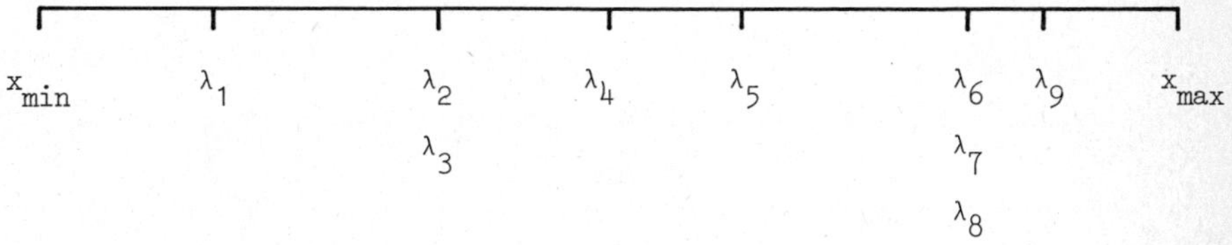

Subsequently polynomial splines will be expressed in terms of B-splines defined upon such a partition. $\lambda_1, \ldots, \lambda_N$ will be termed *interior knots*. In order to define the full set of B-splines required, "*exterior knots*" λ_j, $j < 1$, $j > N$, satisfying

$$\ldots \leqslant \lambda_{-1} \leqslant \lambda_0 = x_{min}, \quad x_{max} = \lambda_{N+1} \leqslant \lambda_{N+2} \leqslant \ldots$$

will be defined. The positions of the knots to be so introduced will not be prescribed at this stage, but a specific choice will be made subsequently and its advantages discussed.

Throughout, prescribed values of x_{min}, x_{max} and $\lambda_1, \ldots, \lambda_N$ will be assumed to satisfy (3).

λ_j is termed a *simple knot* if $\lambda_{j-1} < \lambda_j < \lambda_{j+1}$ and a *knot of multiplicity r* or simply a *coincident* or *multiple knot* if $\lambda_{j-1} < \lambda_j = \ldots = \lambda_{j+r-1} < \lambda_{j+r}$.

The subintervals and the interval index function

Define the *subintervals*

$$I_j = \left\{ x : x \in [\lambda_j, \lambda_{j+1}),\ j = 0, \ldots, N-1; \quad x \in [\lambda_j, \lambda_{j+1}],\ j = N \right\}. \tag{4}$$

Note (the arbitrary decision) that, for $j < N$, interior knot λ_j is associated with subinterval I_j (rather than I_{j+1}). Also note that I_j is *empty* if $\lambda_j = \lambda_{j+1}$.

Define the (piecewise constant) *interval index function*

$$J(x) = k : x \in I_k, \quad x \in [x_{min}, x_{max}].$$

Truncated power functions

The truncated power function (TPF) of *order* n is defined and denoted by

$$(x-\lambda)_+^{n-1} = \left\{(x-\lambda)_+\right\}^{n-1},$$

where

$$(x-\lambda)_+ = \max(x-\lambda,\ 0).$$

For $n > 1$, the TPF of order n is continuous and possesses n-2 continuous derivatives.

Truncated power function representation of polynomial splines

Consider first splines of *order* n (*degree* < n) defined on $[x_{min}, x_{max}]$ with *simple knots* $\lambda_1, \ldots, \lambda_N$. Then s(x) is such a function if for $x \in I_j$, $j = 0, \ldots, N$, it reduces to a polynomial of degree < n, and at λ_j, $j = 1, \ldots, N$, $s^{(r)}(x)$ is continuous for $r = 0, \ldots, n-2$.

Since s(x) can be represented by a polynomial of degree < n in each of the N + 1 subintervals, together with n - 1 continuity conditions at each of the N interior knots it can be described by $n(N+1) - (n-1)N = N + n$ *linear parameters* or *coefficients*. The following derivation of an explicit form for s(x) confirms this result.

Let $p_j(x)$ denote the polynomial to which s(x) reduces for $x \in I_j$. Then $p_1(x)$ can be expressed as $p_0(x) + \delta p_0(x)$, where $\delta p_0(x)$ is an "adjustment polynomial" of degree < n. Now $p_0^{(r)}(x)$ and $p_1^{(r)}(x)$ are to agree at $x = \lambda_1$ for $r = 0, \ldots, n-2$, and hence $\delta p_0^{(r)}(x) = 0$ at λ_1 for $r = 0, \ldots, n-2$. Thus the first n - 1 coefficients in the Taylor representation of $\delta p_0(x)$ about $x = \lambda_1$ must be zero. But since $\delta p_0(x)$ is of degree < n, it follows that $\delta p_0(x) = b_1(x-\lambda_1)^{n-1}$ for some constant b_1, and hence $p_1(x) = p_0(x) + b_1(x-\lambda_1)^{n-1}$. Extension of this argument yields

$$p_j(x) = p_{j-1}(x) + b_j(x-\lambda_j)^{n-1}, \; j = 1, \ldots, N,$$

where each b_j is a constant. It follows that

$$s(x) = p(x) + \sum_{k=1}^{j} b_k(x-\lambda_k)^{n-1}, \; x \in I_j, \; j = 0, \ldots, N,$$

where the subscript has now been dropped from $p_0(x)$. Since $(x-\lambda_j)_+ = 0$, $x < \lambda_j$, this form is identical to

$$s(x) = p(x) + \sum_{j=1}^{N} b_j(x-\lambda_j)_+^{n-1}, \quad x \in [x_{min}, x_{max}].$$

Equivalently,

$$s(x) = \sum_{j=1}^{q} \beta_j\phi_j(x), \quad q = N + n, \tag{5}$$

for some coefficients β_j, where

$$\phi_j(x) = \begin{cases} x^{j-1}, & j = 1, \ldots, n, \\ (x-\lambda_{j-n})_+^{n-1}, & j = n+1, \ldots, q, \end{cases} \tag{6}$$

which portrays clearly that s(x) can be expressed as a linear combination of certain functions. Alternatively, if additional knots λ_j, $j < 0$, satisfying $\ldots < \lambda_{-2} < \lambda_{-1} < \lambda_0$ are specified, the $\phi_j(x)$ can be defined by

$$\phi_j(x) = (x-\lambda_{j-n})_+^{n-1}, \quad j = 1, \ldots, q. \tag{7}$$

(6) and (7) shall be termed *truncated power function bases* for s(x).

A generalization of the above argument enables the restriction that $s^{(r)}(x)$ is continuous for all $r = 0, \ldots, n-2$ to be relaxed at one or more knots. If the value of the interior knot λ_k is repeated exactly ν_k times, this is taken to mean that $s^{(r)}(x)$ there is to be continuous for $r = 0, \ldots, n-1-\nu_k$. The basis is then extended to include functions $(x-\lambda_k)_+^{n-\ell}$, $\ell = 1, \ldots, \nu_k$. Note the general rule that, at a knot, *number of continuity conditions + knot multiplicity = order of spline.* The extended basis now yields a function s(x) containing the correct number of linear parameters with $s^{(r)}(x)$ continuous at $x = \lambda_k$ for $r = 0, \ldots, n-1-\nu_k$.

Inadequacies of the truncated power function basis

Superficially the TPF basis is ideal: the representation (5) is "polynomial-like" and can therefore readily be evaluated, differentiated and integrated - it is only necessary to cope correctly with the "suffix +". Moreover, since (5) is an explicit representation of a spline, it is possible to use the wealth of material available for approximation by functions that are linear in their parameters. For instance spline approximants in various norms - ℓ_1, ℓ_2 and ℓ_∞ - can be computed simply by making use of standard mathematical software (see e.g. Barrodale and Young, 1967).

It will come as no surprise that there are distinct disadvantages in this approach. For instance, the choice of two close knots will give rise to two terms in the TPF basis of the form $(x-\lambda)_+^{n-1}$ and $(x-\lambda-\delta\lambda)_+^{n-1}$, where $\delta\lambda$ is "small". However, if the two knots were coincident, the corresponding terms in the TPF basis would be $(x-\lambda)_+^{n-1}$ and $(x-\lambda)_+^{n-2}$. It would appear reasonable therefore that, for $\delta\lambda$ sufficiently small, the linear combination $d_1(x-\lambda)_+^{n-1} + d_2(x-\lambda)_+^{n-2}$ could be closely approximated, for certain e_1, e_2, by $e_1(x-\lambda)_+^{n-1} + e_2(x-\lambda-\delta\lambda)_+^{n-1}$. However, for "moderate" values of d_1 and d_2, say of order unity, the values of e_1 and e_2 in such an approximation prove to be "large" ($\sim 1/\delta\lambda$), approximately equal and of opposite signs. Thus, even if accurate values for e_1 and e_2 could be computed, loss of significance would occur when a spline containing such terms were evaluated. The fact that the linear multipliers of the TPF representation can exhibit such behaviour means that the TPF basis is *inherently illconditioned*.

A further significant disadvantage of the TPF representation is that its evaluation may be very uneconomical. The worst case occurs when $x \in I_N$, for which $\sim Nn$ multiplications ($\sim N \log_2 n$ with care) are required. One reason for this large amount of work is the "asymmetry" of the TPF basis. If functions of the form $(\lambda-x)_+^{n-1}$ were used instead, most work would then be associated with values of $x \in I_0$ and least with $x \in I_N$.

One final disincentive to the use of the TPF basis must be mentioned. Suppose it is required to determine a spline interpolant to some function or a spline collocant to some ordinary differential equation. In either case it is necessary to construct and solve a system of linear algebraic equations, the coefficient matrix of which contains the values of $\phi_j(x)$ and/or their derivatives at the interpolation or collocation points. The resulting coefficient matrix is lower Hessenberg in form, i.e. its lower triangle is full and some of its leading superdiagonals also certain nonzero elements. This form is a further consequence of the asymmetry of the TPF basis, and the solution of systems having such coefficient matrices demands significantly more computational resources (both time and space) than does the approach discussed in Part II.

A basis is therefore required that is well-conditioned, economical and, ideally, "symmetric" in some sense. Of course, each member of the basis will of necessity be a certain linear combination of the TPF's. The B-splines form such a basis.

The (normalized) B-splines

Let $\ldots, \lambda_{j-n}, \ldots, \lambda_j, \ldots$ be a set of knots with $\lambda_k \leq \lambda_{k+1}$, all k. The (normalized) B-spline $N_{n,j}(x)$ is defined by

$$N_{n,j}(x) = (\lambda_j-\lambda_{j-n})[\lambda_{j-n}, \ldots, \lambda_j](. - x)_+^{n-1}, \tag{8}$$

if $\lambda_{j-n} < \lambda_j$, and the zero function otherwise.

Note that if an operation is to be taken with respect to one of several variables of a function, the relevant variable is indicated by the *placeholder* notation. For instance, $[\lambda_{j-1}, \lambda_j]f(., x)$ denotes the divided difference at λ_{j-1}, λ_j of the bivariate function f with respect to its first variable, viz. the value of

$$\frac{f(\lambda_{j-1}, x) - f(\lambda_j, x)}{\lambda_{j-1} - \lambda_j} .$$

Example Suppose $\lambda_{j-2} < \lambda_{j-1} < \lambda_j$. Then, using (1),

$$\begin{aligned} N_{2,j}(x) &= (\lambda_j-\lambda_{j-2})[\lambda_{j-2}, \lambda_{j-1}, \lambda_j](. - x)_+ \\ &= \frac{(\lambda_j-x)_+ - (\lambda_{j-1}-x)_+}{\lambda_j - \lambda_{j-1}} - \frac{(\lambda_{j-1}-x)_+ - (\lambda_{j-2}-x)_+}{\lambda_{j-1} - \lambda_{j-2}} \\ &= \frac{(\lambda_{j-2}-x)_+}{\lambda_{j-1} - \lambda_{j-2}} - \left(\frac{1}{\lambda_{j-1} - \lambda_{j-2}} + \frac{1}{\lambda_j - \lambda_{j-1}} \right) (\lambda_{j-1}-x)_+ + \frac{(\lambda_j-x)_+}{\lambda_j - \lambda_{j-1}} , \end{aligned}$$

which has the appearance:

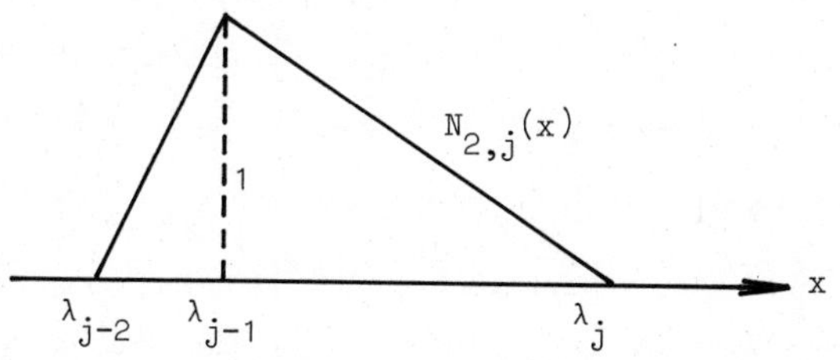

Example Now suppose $\lambda_{j-2} = \lambda_{j-1} < \lambda_j$. Then, again using (1),

$$N_{2,j}(x) = \frac{(\lambda_j-x)_+ - (\lambda_{j-1}-x)_+}{\lambda_j - \lambda_{j-1}} - (\lambda_{j-1}-x)_+^0 ,$$

which looks like:

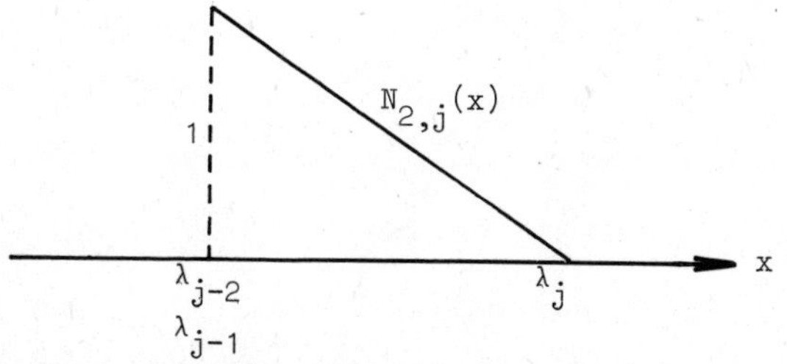

Basic recurrence relation for normalized B-splines

The normalized B-splines satisfy the recurrence relation

$$N_{n,j}(x) = \left(\frac{x - \lambda_{j-n}}{\lambda_{j-1} - \lambda_{j-n}}\right) N_{n-1,j-1}(x) + \left(\frac{\lambda_j - x}{\lambda_j - \lambda_{j-n+1}}\right) N_{n-1,j}(x), \; n > 1, \tag{9}$$

with

$$N_{1,j}(x) = \begin{cases} 1, & x \in I_{j-1}, \\ 0, & x \notin I_{j-1}. \end{cases} \tag{10}$$

Proof (10) follows immediately from (8). To obtain (9) write

$(\lambda-x)_+^{n-1} = (\lambda-x)(\lambda-x)_+^{n-2}$ and use (2) to give

$$[\lambda_{j-n}, \ldots, \lambda_j](.-x)_+^{n-1} =$$

$$= \sum_{k=j-n}^{j} ([\lambda_{j-n}, \ldots, \lambda_k](.-x))([\lambda_k, \ldots, \lambda_j](.-x)_+^{n-2})$$

$$= (\lambda_{j-n}-x)[\lambda_{j-n}, \ldots, \lambda_j](.-x)_+^{n-2} + (1)[\lambda_{j-n+1}, \ldots, \lambda_j](.-x)_+^{n-2},$$

since divided differences of order > 2 of linear polynomials are zero. Thus, since from (1),

$$[\lambda_{j-n}, \ldots, \lambda_j]f = ([\lambda_{j-n+1}, \ldots, \lambda_j]f - [\lambda_{j-n}, \ldots, \lambda_{j-1}]f)/(\lambda_j-\lambda_{j-n}),$$

it follows that

$$[\lambda_{j-n}, \ldots, \lambda_j](.-x)_+^{n-1} =$$

$$= \left(\frac{x - \lambda_{j-n}}{\lambda_j - \lambda_{j-n}}\right)[\lambda_{j-n}, \ldots, \lambda_{j-1}](.-x)_+^{n-2} + \left(\frac{\lambda_j - x}{\lambda_j - \lambda_{j-n}}\right)[\lambda_{j-n+1}, \ldots, \lambda_j](.-x)_+^{n-2},$$

from which, by use of (8), the result (9) follows.

The unnormalized B-splines

The unnormalized B-splines are defined by

$$M_{n,j}(x) = N_{n,j}(x)/(\lambda_j-\lambda_{j-n})$$

and therefore satisfy the recurrence relation

$$M_{n,j}(x) = \frac{(x-\lambda_{j-n})M_{n-1,j-1}(x) + (\lambda_j-x)M_{n-1,j}(x)}{\lambda_j - \lambda_{j-n}}, \quad n > 1,$$

with

$$M_{1,j}(x) = \begin{cases} (\lambda_j-\lambda_{j-1})^{-1}, & x \in I_{j-1}, \\ 0, & x \notin I_{j-1}. \end{cases}$$

B-spline properties

Positivity and compact support

$$N_{n,j}(x) > 0, \quad x \in (\lambda_{j-n}, \lambda_j),$$

$$N_{n,j}(x) = 0, \quad x \notin [\lambda_{j-n}, \lambda_j]. \tag{11}$$

These results follow immediately from (9) and (10).

Normalization

$$\sum_j N_{n,j}(x) \equiv 1, \quad x \in [x_{min}, x_{max}], \tag{12}$$

$$\int_{x_{min}}^{x_{max}} M_{n,j}(x)dx = n^{-1}. \tag{13}$$

Proof of (12). From (9) and (10),

$$\sum_j N_{n,j}(x) = \sum_j \left(\frac{x - \lambda_{j-n}}{\lambda_{j-1} - \lambda_{j-n}} \right) N_{n-1,j-1}(x) + \sum_j \left(\frac{\lambda_j - x}{\lambda_j - \lambda_{j-n+1}} \right) N_{n-1,j}(x)$$

$$= \sum_j \left(\frac{x - \lambda_{j-n+1}}{\lambda_j - \lambda_{j-n+1}} \right) N_{n-1,j}(x) + \sum_j \left(\frac{\lambda_j - x}{\lambda_j - \lambda_{j-n+1}} \right) N_{n-1,j}(x)$$

$$= \sum_j N_{n-1,j}(x) = \sum_j N_{n-2,j}(x) = \dots = \sum_j N_{1,j}(x) = 1.$$

Proof of (13). This result follows immediately as a special case of the indefinite integral of a B-spline, considered later.

Numerical evaluation of the B-splines

Given x_{min}, x_{max}, $\lambda_1, \ldots, \lambda_N$ and n, all $N_{n,j}(x)$ that are nonzero for a specified value of x may be evaluated as follows. The interval index $J = J(x)$ is first determined; then $x \in I_J$. Since the support of $N_{r,k}(x)$ is $[\lambda_{k-r}, \lambda_k)$, the only B-splines of orders 1, ..., n that are possibly nonzero in I_J are (omitting the argument for brevity) those in the *B-spline table*

$$\begin{array}{ccccc}
 & & & \cdot & N_{n,J+1} \\
 & & & \cdot & \\
 & & N_{3,J+1} & \cdot & N_{n,J+2} \\
 & & & \cdot & \cdot \\
 & N_{2,J+1} & & \cdot & \cdot \\
N_{1,J+1} & & N_{3,J+2} & \cdot & \cdot \\
 & & & \cdot & \cdot \\
 & N_{2,J+2} & & \cdot & \cdot \\
 & & & \cdot & \cdot \\
 & & N_{3,J+3} & \cdot & \cdot \\
 & & & & N_{n,J+n}
\end{array}$$

The table can be generated column by column using (9) and (10) and the fact that uncited entries in the table are to be taken as zero.

About $\frac{3}{2}n^2$ operations (here and subsequently one operation denotes one multiplication or one division) are required to form all entries in the table, i.e. to evaluate all B-splines of all orders from 1 to n that are nonzero in I_J. The scheme is unconditionally stable for any knot set (Cox, 1972):

$$|\bar{N}_{n,j}(x) - N_{n,j}(x)| \lesssim L_1(n)\eta N_{n,j}(x), \tag{14}$$

where here and subsequently a bar is used to denote a computed quantity, $L_k(n)$, for any subscript k, is a linear function of n, and η is the *unit roundoff* (the smallest machine representable number such that the *computed value* of $1 + \eta$ exceeds unity). The fundamental reason for the stability of the scheme is that each stage of the process involves taking a nonnegative combination of nonnegative quantities, and thus no loss of significance through cancellation can occur.

Recurrence relations for the B-spline derivatives

The following two recurrence relations hold for the derivatives of normalized B-splines:

$$N_{n,j}^{(\ell)}(x) = (n-1)\left[\frac{N_{n-1,j-1}^{(\ell-1)}(x)}{\lambda_{j-1}-\lambda_{j-n}} - \frac{N_{n-1,j}^{(\ell-1)}(x)}{\lambda_j - \lambda_{j-n+1}}\right], \tag{15}$$

$$N_{n,j}^{(\ell)}(x) = \frac{n-1}{n-\ell-1}\left[\left(\frac{x-\lambda_{j-n}}{\lambda_{j-1}-\lambda_{j-n}}\right)N_{n-1,j-1}^{(\ell)}(x)+\left(\frac{\lambda_j - x}{\lambda_j-\lambda_{j-n+1}}\right)N_{n-1,j}^{(\ell)}(x)\right]. \tag{16}$$

Proof of (15). The use of (1), (8) and

$$\frac{d}{dx}(\lambda-x)_+^{n-1} = -(n-1)(\lambda-x)_+^{n-2}$$

gives

$$N'_{n,j}(x) = -(n-1)([\lambda_{j-n+1}, \ldots, \lambda_j](.-x)_+^{n-2}$$

$$- [\lambda_{j-n}, \ldots, \lambda_{j-1}](.-x)_+^{n-2})$$

$$= -(n-1)\left(\frac{N_{n-1,j}(x)}{\lambda_j - \lambda_{j-n+1}} - \frac{N_{n-1,j-1}(x)}{\lambda_{j-1}-\lambda_{j-n}}\right),$$

the $(\ell-1)$-fold differentiation of which yields (15).

Proof of (16). This result follows by induction on ℓ together with the use of (9) and (15).

To determine all $N_{n,j}^{(r)}(x)$ that are nonzero for specified n, r and x, the first n-r columns of the *B-spline derivative table* may be formed using (9) and then the final r columns using (15).

The choice of exterior knots

For a given interval $[x_{min}, x_{max}]$, choose exterior knots

$$\lambda_j = \begin{cases} x_{min}, & j < 0, \\ x_{max}, & j > N+1. \end{cases} \tag{17}$$

Reasons for this choice, which will be adopted throughout, are given in Part II of this paper.

B-spline values and derivatives at interval endpoints

Values and derivatives of the B-splines at the interval endpoints are required when determining interpolating and approximating splines that satisfy specified boundary conditions (see Part II). From (15) and the compact support of the B-splines, the following results are readily obtained (Cox, 1975a):

$$\mathrm{sign}\,(N_{n,j}^{(\ell)}(x_{min})) = \begin{cases} (-1)^{\ell-j+1}, & j \leq \ell + 1 \\ 0, & j > \ell + 1, \end{cases}$$

$$\mathrm{sign}\,(N_{n,j}^{(\ell)}(x_{max})) = \begin{cases} (-1)^{N+n-j}, & j \geq N + n - \ell, \\ 0, & j < N + n - \ell. \end{cases} \tag{18}$$

It follows from (18) and (12) that

$$N_{n,j}(x_{min}) = N_{n,N+n+1-j}(x_{max}) = \begin{cases} 1, & j = 1, \\ 0, & j > 1. \end{cases}$$

Moreover, it can be established (Cox, 1975a) that the application of (15) gives

$$|\bar{N}_{n,j}^{(\ell)}(x) - N_{n,j}^{(\ell)}(x)| \leq L_2(n)\eta|N_{n,j}^{(\ell)}(x)|, \quad x = x_{min}, x_{max}. \tag{19}$$

The main reason for such a good bound is that (18) implies that contributory terms always reinforce one another in the use of (15) at $x = x_{min}$ and $x = x_{max}$ and hence no loss of significance can occur.

Indefinite integrals of B-splines

For $x \in [x_{min}, x_{max}]$,

$$\int_{-\infty}^{x} N_{n,j}(t)dt = \begin{cases} 0 & , \quad x < \lambda_{j-n}, \\ \dfrac{\lambda_j - \lambda_{j-n}}{n} \displaystyle\sum_{k=j+1}^{j+n} N_{n+1,k}(x) & , \quad \lambda_{j-n} \leq x < \lambda_j, \\ \dfrac{\lambda_j - \lambda_{j-n}}{n} & , \quad \lambda_j \leq x. \end{cases}$$

Proof The result is obtained following integration of (15) with $\ell = 1$ together with the compact support property (11).

Butterfield (1976) has shown that recurrence relation (16) also holds for all $\ell < 0$, in which case $N_{n,j}^{(\ell)}(x)$ is to be interpreted as the $(-\ell)$-fold indefinite integral of $N_{n,j}(x)$.

B-spline representation of polynomial splines

The B-splines $N_{n,j}(x)$, $j = 1, \ldots, q$, form a basis for splines of order n on $[x_{min}, x_{max}]$ with knots $\lambda_1, \ldots, \lambda_N$ (Curry and Schoenberg, 1966). Thus if s(x) is such a spline, it has the *B-spline representation*

$$s(x) = \sum_{j=1}^{q} c_j N_{n,j}(x), \quad x \in [x_{min}, x_{max}]. \tag{20}$$

The c_j are termed *B-spline coefficients*. From the compact support property,

$$s(x) = \sum_{j=J+1}^{J+n} c_j N_{n,j}(x), \quad x \in I_J, \quad J \in \{0, N\}. \tag{21}$$

Properties of the B-spline representation

Convex combinations

From the normalization and positivity properties of B-splines, s(x) is a *convex combination* of $c_1, \ldots, c_q$. In particular, if $x \in I_J$, s(x) is a convex combination of $c_{J+1}, \ldots, c_{J+n}$.

Local bounds

It follows, in particular, that

$$\min c_j \leqslant s(x) \leqslant \max c_j, \; x \in I_J, \tag{22}$$

where the min and max are taken over $j = J+1, \ldots, J+n$. This is a *local* result, i.e. it depends only upon the coefficients of the B-splines that are nonzero in I_J. It will be seen subsequently that a generalization of this result applies also to the derivatives of s(x).

Numerical evaluation of the B-spline representation

Two schemes, A and B, are given for the evaluation of (20) for a specified value of x. Each scheme consists of two stages, the first of which is the determination of J such that $x \in I_J$ using binary search, which requires $\sim \log_2 N$ comparisons. The second stage involves the evaluation of (21) which can be undertaken in two ways. The most obvious way, Scheme A, is to form the basis values $N_{n,j}(x)$, $j = J+1, \ldots, J+n$, to multiply them by the respective B-spline coefficients and sum. Scheme B is based upon the following considerations.

Use of (9) for the specified value of x (de Boor, 1972) gives

$$s(x) = \sum_{j=J+1}^{J+n} c_j \left(\frac{(x-\lambda_{j-n})N_{n-1,j-1}(x)}{\lambda_{j-1} - \lambda_{j-n}} + \frac{(\lambda_j - x)N_{n-1,j}(x)}{\lambda_j - \lambda_{j-n+1}} \right)$$

$$= \sum_{j=J}^{J+n-1} c_{j+1} \left(\frac{x - \lambda_{j-n+1}}{\lambda_j - \lambda_{j-n+1}} \right) N_{n-1,j}(x) + \sum_{j=J+1}^{J+n} c_j \left(\frac{\lambda_j - x}{\lambda_j - \lambda_{j-n+1}} \right) N_{n-1,j}(x).$$

But, from the compact support property, $N_{n-1,J}(x) = N_{n-1,J+n}(x) = 0$ for $x \in I_J$. Thus

$$s(x) = \sum_{j=J+1}^{J+n-1} \left(\frac{(x-\lambda_{j-n+1})c_{j+1} + (\lambda_j - x)c_j}{\lambda_j - \lambda_{j-n+1}} \right) N_{n-1,j}(x)$$

$$= \sum_{j=J+1}^{J+n-1} c_j^{[1]} N_{n-1,j}(x), \quad c_j^{[1]} = \frac{(x-\lambda_{j-n+1})c_{j+1} + (\lambda_j - x)c_j}{\lambda_j - \lambda_{j-n+1}}.$$

Repetition of this process a further n-1 times yields, using (10),

$$s(x) = c_{J+1}^{[n]} N_{1,J+1}(x) = c_{J+1}^{[n]},$$

where

$$c_j^{[\ell]} = \begin{cases} c_j & \ell = 0, \\ \dfrac{(x-\lambda_{j-n+\ell})c_{j+1}^{[\ell-1]} + (\lambda_j - x)c_j^{[\ell-1]}}{\lambda_j - \lambda_{j-n+\ell}}, & \ell = 1, \ldots, n. \end{cases}$$

Note that $c_j^{[\ell]}$, $\ell > 0$, is a *convex combination* of $c_j^{[\ell-1]}$ and $c_{j+1}^{[\ell-1]}$.

If n = 4, for example, the numbers in the following array are computed column by column:

$$
\begin{array}{lllll}
c_{J+1} = c_{J+1}^{[0]} & & & & \\
 & c_{J+1}^{[1]} & & & \\
c_{J+2} = c_{J+2}^{[0]} & & c_{J+1}^{[2]} & & \\
 & c_{J+2}^{[1]} & & c_{J+1}^{[3]} & = s(x) \\
c_{J+3} = c_{J+3}^{[0]} & & c_{J+2}^{[2]} & & \\
 & c_{J+3}^{[1]} & & & \\
c_{J+4} = c_{J+4}^{[0]} & & & &
\end{array}
$$

Both schemes take about $\frac{3}{2}n^2$ operations, the second being marginally faster. The schemes have comparable numerical stability (Cox, 1978c): each provides a computed value $\bar{s}(x)$ satisfying

$$|\bar{s}(x) - s(x)| \leqslant L_3(n)\eta \max_{J+1 \leqslant j \leqslant J+n} |c_j|.$$

Finally, Scheme A is *significantly* faster if several splines based upon the same knot set are to be evaluated at a specified x-value (as in computations with tensor product splines). If there are r such splines, Scheme A takes about $\frac{3}{2}n^2 + nr$ operations, whereas Scheme B takes about $\frac{3}{2}n^2 r$ operations.

Derivatives of the B-spline representation

Differentiation of (20) and the use of (15) gives

$$s'(x) = \sum_{j=1}^{q} c_j N'_{n,j}(x) = \sum_{j=1}^{q} (n-1)c_j \left(\frac{N_{n-1,j-1}(x)}{\lambda_{j-1} - \lambda_{j-n}} - \frac{N_{n-1,j}(x)}{\lambda_j - \lambda_{j-n+1}} \right).$$

Thus, since $N_{n-1,0}(x) = N_{n-1,q}(x) = 0$ for $x \in [x_{min}, x_{max}]$,

$$s'(x) = \sum_{j=1}^{q-1} (n-1) \left(\frac{c_{j+1} - c_j}{\lambda_j - \lambda_{j-n+1}} \right) N_{n-1,j}(x).$$

The process can be repeated. If

$$s^{(\ell)}(x) = \sum_{j=1}^{q-\ell} c_j^{(\ell)} N_{n-\ell,j}(x),$$

then

$$c_j^{(\ell)} = \begin{cases} c_j, & \ell = 0, \\ (n-\ell)\left[\dfrac{c_{j+1}^{(\ell-1)} - c_j^{(\ell-1)}}{\lambda_j - \lambda_{j-n+\ell}}\right], & \ell = 1, \ldots, n-1. \end{cases} \tag{23}$$

Note that if $\lambda_j = \lambda_{j-n+\ell}$, $c_j^{(\ell)}$ cannot be computed from (23). However, since $N_{n-\ell,j}(x)$ is defined to be the zero function when this condition obtains, an arbitrary value can be assigned to $c_j^{(\ell)}$, since its value will not influence that of $s^{(\ell)}(x)$.

Bounds for the derivatives

Just as (22) provides local bounds for values of $s(x)$, the $c_j^{(\ell)}$ provide local bounds for its derivatives:

$$\min c_j^{(\ell)} \leqslant s^{(\ell)}(x) \leqslant \max c_j^{(\ell)}, \quad x \in I_J,$$

where the min and max are taken over $j = J+1, \ldots, J+n-\ell$.

Indefinite integrals of the B-spline representation

The ℓ-fold indefinite integral $s^{(-\ell)}(x)$ of (20) is given by

$$s^{(-\ell)}(x) = \sum_{j=1}^{q+\ell} c_j^{(-\ell)} N_{n+\ell,j}(x) + \sum_{k=1}^{\ell} \frac{a_k}{(\ell-k)!} x^{\ell-k}, \tag{24}$$

where the $c_j^{(-\ell)}$ are defined recursively by

$$c_j^{(-\ell)} = \begin{cases} 0 & , \quad j \leqslant \ell, \\ c_{j-1}^{(-\ell)} + \left[\dfrac{\lambda_{j-1} - \lambda_{j-n-\ell}}{n+\ell-1}\right] c_{j-1}^{(1-\ell)} , & j > \ell, \end{cases}$$

and $a_1, \ldots, a_\ell$ are arbitrary constants of integration.

Proof A single differentiation of the right hand side of (24), after the inclusion of an arbitrary additive constant, yields, through the application of (23), an expression of similar form with ℓ replaced by $\ell-1$.

It may be inconvenient in some applications to work with the expression (24) since it involves both B-splines and powers of x. In such cases the power series part may be converted in a stable way to B-spline form using the algorithm given in Cox(1978a). Alternatively, immediately after each integration the corresponding constant of integration may be added, by virtue of (12), to each newly produced B-spline coefficient.

Summary

Part I of this paper has introduced polynomial splines, and discussed B-splines and the representation of splines in terms of B-splines. Various valuable properties of B-splines and B-spline representations have been derived and discussed. Part II makes use of the B-spline representation and its properties to construct stable and efficient schemes for spline interpolation and least squares spline approximation.

PART II: SPLINE INTERPOLATION AND LEAST SQUARES SPLINE APPROXIMATION

Derivative-free spline interpolation

Suppose the following information is prescribed: an interval $[x_{min}, x_{max}]$; data points (x_i, f_i), $i = 1, \ldots, m$, with $x_{min} \leq x_1 < \ldots < x_m \leq x_{max}$; an integer n, $0 < n \leq m$; $N = m - n$ interior knots $\lambda_1, \ldots, \lambda_N$, with $x_{min} < \lambda_1 \leq \ldots \leq \lambda_N < x_{max}$ and $\lambda_{j-n} < \lambda_j$ for all j. (The last condition ensures that all B-splines are non-null.) The spline interpolation problem is to determine a spline $s(x)$ of order n with knots $\lambda_1, \ldots, \lambda_N$ satisfying

$$s(x_i) = f_i, \quad i = 1, \ldots, m.$$

If this problem has a solution, it satisfies, through I(20), with $q = m$, the linear equations

$$\sum_{j=1}^{m} c_j N_{n,j}(x_i) = f_i, \quad i = 1, \ldots, m.$$

In matrix form these equations are

$$Ac = f, \tag{1}$$

where $a_{ij} = N_{n,j}(x_i)$, $c = (c_1, \ldots, c_m)^T$, $f = (f_1, \ldots, f_m)^T$.

Conditions for solution

If A is invertible, a solution to this interpolation problem exists and is unique. Necessary and sufficient conditions for A to be invertible were first established by Schoenberg and Whitney (1953). (For a more elementary proof see de Boor (1976)). These conditions, which shall be termed here *conditions SW*, are

$$\lambda_{j-n} < x_j < \lambda_j, \; j = 1, \ldots, m. \tag{2}$$

In terms of the B-splines these conditions state that, for $j = 1, \ldots, m$, the jth data abscissa lies strictly within the support of the jth B-spline, and, in terms of the matrix formulation, each diagonal element a_{jj}, $j = 1, \ldots, m$, is nonzero. It may readily be verified that for the choice I(17) of exterior knots, the left hand inequality in (2) may be relaxed to include equality when $j = 1$ and the right hand one similarly when $j = m$.

Solution of the defining equations

System (1) can be formed stably, because of I(14), and is well balanced, each row sum being unity, as a result of I(12). Moreover, it has a special form, advantage of which can be taken to obtain its solution particularly efficiently. Because of the compact support property I(11), each row of A contains (at most) n nonzero elements, and these occur in contiguous positions in the row. Moreover, the ordering of the data implies that the column index p_i, say, of the *first* nonzero element in row i is a nondecreasing function of i. Note that $p_i = J_i + 1$, where J_i is the integer such that $x_i \in I_{J_i}$ (see I(4) for the definition of I_{J_i}). Let v_j be the row index of the *last* nonzero element in column j of A. It is easily verified that

$$v_j = \max k : p_k \leq j.$$

Standard Gaussian elimination for reducing (1) to upper triangular form, without taking account of structure, can be described compactly by the algorithm

```
for j := 1, ..., n
    for i := j + 1, ..., n
        subtract multiple of row j of (A, f) from row i
           of (A, f) to eliminate element (i, j) of A
```

in which A and f denote the "partially eliminated" A and f, the outer loop counts over rows used as pivotal rows and the inner loop over rows modified by the jth pivotal row. The single change necessary to adapt this algorithm to the special structure of A is to limit the inner loop to values of i for which element (i, j) is nonzero, i.e. to replace the second for-statement by "for $i = j + 1, \ldots, v_j$". This variant of Gaussian elimination is very efficient: fewer than mn^2 operations are required, an "average" number being $\frac{1}{2}mn^2$ (Cox, 1975b). Backsubstitution for c requires a further mn operations. Since about $\frac{3}{2}n^2$ operations are required to form each row of A, the complete process requires about $2mn^2$ operations.

Note that pivoting is unnecessary. The matrix A is totally positive, i.e. all its minors are nonnegative (Karlin, 1968): a linear system with such a coefficient matrix can be solved quite safely without pivoting (de Boor and Pinkus, 1977). Note also that the elimination can be carried out *in situ* and thus no additional array space to accommodate fill-in is required.

Spline interpolation with derivative boundary conditions

The following spline interpolation problem is now considered. In addition to an interval $[x_{min}, x_{max}]$ and data points (x_i, f_i), $i = 1, \ldots, m$, with $x_{min} < x_1 < x_2 < \ldots < x_m < x_{max}$, suppose that boundary values $f^{(r)}(x_{min})$, $r = 0, \ldots, m_1 - 1$, and $f^{(r)}(x_{max})$, $r = 0, \ldots, m_2 - 1$, with $m_1 \leqslant n$, $m_2 \leqslant n$, are specified. Also suppose that an integer n, $0 < n \leqslant \tilde{m} = m + m_1 + m_2$, and $N = \tilde{m} - n$ interior knots $\lambda_1, \ldots, \lambda_N$, with $x_{min} < \lambda_1 \leqslant \lambda_2 \leqslant \ldots \leqslant \lambda_N < x_{max}$ and $\lambda_{j-n} < \lambda_j$ for all j, are prescribed.

The problem is to determine a spline interpolant $s(x)$ of order n with knots $\lambda_1, \ldots, \lambda_N$ satisfying

$$\begin{aligned} s(x_i) &= f_i, && i = 1, \ldots, m, \\ s^{(r)}(x_{min}) &= f^{(r)}(x_{min}), && r = 0, \ldots, m_1 - 1, \\ s^{(r)}(x_{max}) &= f^{(r)}(x_{max}), && r = 0, \ldots, m_2 - 1. \end{aligned}$$

If this problem has a solution it satisfies the linear equations

$$\begin{aligned} \sum_{j=1}^{q} c_j N_{n,j}^{(r)}(x_{min}) &= f^{(r)}(x_{min}), \quad r = 0, \ldots, m_1 - 1, \\ \sum_{j=1}^{q} c_j N_{n,j}(x_i) &= f_i, \quad i = 1, \ldots, m, \\ \sum_{j=1}^{q} c_j N_{n,j}^{(r)}(x_{max}) &= f^{(r)}(x_{max}), \quad r = m_2 - 1, m_2 - 2, \ldots, 0, \end{aligned} \tag{3}$$

where the ordering of the equations is suggestive as regards structure of their associated matrix. In terms of the matrix form given in (1), $a_{ij} = N_{n,j}^{(i-1)}(x_{min})$, $i = 1, \ldots, m_1$; $= N_{n,j}(x_{i-m_1})$, $i = m_1 + 1, \ldots, m_1 + m$; $= N_{n,j}^{(\tilde{m}-i)}(x_{max})$, $i = m_1 + m + 1, \ldots, \tilde{m}$; $f_i = f^{(i-1)}(x_{min})$, $i = 1, \ldots, m_1$; $= f_{i-m_1}$, $i = m_1 + 1, \ldots, m_1 + m$; $= f^{(\tilde{m}-i)}(x_{max})$, $i = m_1 + m + 1, \ldots, \tilde{m}$.

Conditions for solution

By regarding derivative end conditions as a limiting case of function values at a set of distinct points, conditions SW for this problem become

$$\lambda_{j-n} < u_j < \lambda_j, \; j = 1, \ldots, \tilde{m},$$

where

$$(u_1, \ldots, u_{\tilde{m}}) = (\underbrace{x_{min}, \ldots, x_{min}}_{m_1}, x_1, \ldots, x_m, \underbrace{x_{max}, \ldots, x_{max}}_{m_2}).$$

A is invertible and a solution therefore exists and is unique if and only if these conditions are satisfied.

Solution of the defining equations

Again, the linear system (1) based on (3) can be formed stably, because of I(14) and I(19), and each row of A has at most n contiguous nonzero elements. As a consequence of I(18) the first m_1 rows are lower triangular in form and the last m_2 rows upper triangular. Thus the first m_1 and last m_2 variables can be eliminated immediately. By a simple extension of the results of Kozak (1980) the remaining matrix is totally positive and thus the corresponding equations can again be solved safely by Gaussian elimination without pivoting. This process is equivalent to the initial determination of the first and last few B-spline coefficients $c_1, \ldots, c_{m_1}, c_{\tilde{m}-m_2+1}, \ldots, c_{\tilde{m}}$, followed by the construction of the interpolant $\tilde{s}(x) = s(x) - \delta s(x)$, with

$$\delta s(x) = \left(\sum_{j=1}^{m_1} + \sum_{j=m_1+m+1}^{\tilde{m}} \right) c_j N_{n,j}(x),$$ to the modified data $\tilde{f}_i = f_i - \delta s(x_i)$,

$i = 1, \ldots, m$. Note that compact support I(11) implies that only values of f_i for which x_i is contained within the first m_1 or the last m_2 subintervals need to be so modified. See Cox (1978b) for further details.

Least squares spline approximation

The following information is assumed to be prescribed: an interval $[x_{min}, x_{max}]$; data points (x_i, f_i), $i = 1, \ldots, m$, with $x_{min} \leq x_i \leq x_{max}$ for all i; positive weights $w_1, \ldots, w_m$; an integer $n > 0$, N (≥ 0) interior knots $\lambda_1, \ldots, \lambda_N$, with $x_{min} < \lambda_1 \leq \ldots \leq \lambda_N < x_{max}$. The least squares spline approximation problem is to determine a spline s(x) of order n with knots $\lambda_1, \ldots, \lambda_N$ that satisfies the *observation equations*

$$s(x_i) = f_i, \quad i = 1, \ldots, m,$$

in a weighted least squares sense. That is, coefficients $c = (c_1, \ldots, c_q)^T$, $q = N + n$, in the B-spline representation I(20) are sought such that the weighted residual sum of squares $\sum_{i=1}^{m} e_i^2$, where

$$e_i = w_i(f_i - s(x_i))$$

is minimized.

The information specifying this problem should be compared with that for the derivative-free spline interpolation problem and the following points noted. The x_i are no longer restricted to be ordered, but the algorithm described below will be particularly efficient if they are. The w_i would normally be chosen inversely proportional to the estimated standard deviation of the error in f_i. If the f_i are all specified to comparable accuracy, the w_i may all be set to unity. N is normally no longer equal to m - n and is typically much smaller. The restriction that $\lambda_{j-n} < \lambda_j$ for all j is no longer essential for the approach described, although it would normally be preferable to adhere to it.

In matrix terms, the problem becomes

$$\min_c \|e\|_2^2 = e^T e, \tag{4}$$

where

$$e = W(f - Ac). \tag{5}$$

Here $e = (e_1, \ldots, e_m)^T$, $f = (f_1, \ldots, f_m)^T$, W is the diagonal matrix of order m with diagonal elements w_i, and A is the m by q rectangular matrix with $a_{ij} = N_{n,j}(x_i)$.

Conditions for solution

Every linear least squares problem admits solution and thus a least squares spline approximant always exists. The solution satisfies the *normal equations*

$$A^T W^2 Ac = A^T W^2 f \tag{6}$$

and is unique if and only if $A^T W^2 A$ is invertible, i.e. A of full rank. This will be the case if and only if an ordered subset of the x_i satisfies the Schoenberg-Whitney conditions, i.e. if there exists $\{u_1, \ldots, u_q\} \subset \{x_1, \ldots, x_m\}$, with $u_1 < \ldots <_q$, such that

$$\lambda_{j-n} < u_j < \lambda_j, \quad j = 1, \ldots, q. \tag{7}$$

We shall say that conditions SW are satisfied or violated according to whether or not such a subset exists. The following algorithm determines whether the solution is unique in the case where the x_i are nondecreasing:

Set u_0 equal to any value smaller than x_1.

For $j = 1, \ldots, q$, select $u_j = \min_i x_i : x_i > u_{j-1}$ and $x_i \in (\lambda_{j-n}, \lambda_j)$.

The solution is unique if and only if all q steps can be completed.

Solution of the defining equation

Solution of (6) is most reliably effected by orthogonal triangularization of WA, since explicit formation of the normal equations worsens the conditioning of the problem. Let $y = Wf$ and $X = WA$. Then (5) becomes

$$e = y - Xc.$$

Let Q be an orthogonal matrix ($Q^TQ = I$) of order m determined such that

$$X = Q\begin{pmatrix} R \\ 0 \end{pmatrix},$$

where R is an upper triangular matrix of order q. Also let

$$y = Q\theta = Q\begin{pmatrix} \theta_1 \\ \theta_2 \end{pmatrix},$$

where θ_1 has q elements. Then

$$e = Q\begin{pmatrix} \theta_1 - Rc \\ \theta_2 \end{pmatrix},$$

giving

$$e^Te = \|\theta_1 - Rc\|_2^2 + \|\theta_2\|_2^2 ,$$

since Q is orthogonal. e^Te in (4) is thus minimized, with value $\|\theta_2\|_2^2$, when c satisfies the triangular system

$$Rc = \theta_1. \tag{8}$$

If A is of full rank then so is R and (8) may be solved immediately by backsubstitution. Strategies for treating rank deficient cases are discussed subsequently.

A detailed description of an algorithm based upon the use of Givens rotations for determining R and θ is given in Cox (1981). The method employed, which takes full account of structure, processes the rows of A one at a time, so the storage requirement for A is reduced to that of only one of its rows. Total storage requirements are about $nq + 2n$ floating point words, the operation count is about mn^2 if the rows of A correspond to ordered x_i and about mqn if they correspond to randomly disordered x_i. Note that in the former case the operation count is essentially independent of N, and that the time penalty resulting from disordered data is a factor of order q/n.

Nonuniqueness in least squares spline approximation problems

We distinguish between three possibilities that can arise during the solution of a least squares spline approximation problem:

(i) conditions SW are satisfied and the triangular system of equations (8) is well conditioned,

(ii) conditions SW are satisfied and (8) is not well conditioned,

(iii) conditions SW are violated and (hence) (8) is not invertible.

If (i) obtains, (8) defines a stable solution to the problem. Although mathematically the remaining two possibilities are quite different from each other in character, numerically the differences may be undetectable.

To see this, consider the following simple example with $m > n$, $N = 1$, $x_{min} = -1$, $x_{max} = 1$, $\lambda_1 = 0$, unit weights, $m - 1$ well-distributed, distinct data points in $[x_{min}, \lambda_1)$, and one in $(\lambda_1, x_{max}]$ at $x = \delta > 0$. The matrix of B-spline values in the case $m = 6$, $n = 4$ has the form

$$A = \begin{bmatrix} \times & \times & \times & \times & \\ \times & \times & \times & \times & \\ \times & \times & \times & \times & \\ \times & \times & \times & \times & \\ \times & \times & \times & \times & \\ & \times & \times & \times & \times \end{bmatrix}. \tag{9}$$

It is readily verified that the bottom right element $a_{m,n+1} = N_{n,n+1}(\delta) = \delta^{n-1}$. Let ε be a computer threshold value set such that values smaller than ε in absolute value are regarded as zero. Then A would be judged rank deficient if $\delta^{n-1} < \varepsilon$. If, for example, $n = 4$ (cubic splines) and $\varepsilon = 10^{-7}$ (approximately the unit roundoff of some computers), a value of $\delta < 0.0046$ would cause A to be so regarded. Note further that if δ falls below the $(n-1)^{st}$ root of the machine's underflow threshold, the *stored* matrix will be rank deficient regardless of the value of ε. Thus data requiring a perturbation of δ to cause conditions SW to be violated *can* result in a matrix that is within δ^{n-1} of being singular. We are unaware of a *general* relationship between the degree of satisfaction of conditions SW and the closeness of A to singularity.

We firmly believe therefore that algorithms for least squares spline approximation problems should not simply use conditions SW as the sole criterion for whether a problem is well-posed or not. Conditions SW should certainly wherever possible be checked before processing the data (it will not be possible if the data is disordered), since doing so may promptly identify some problems for which grossly incorrect data has inadvertently been provided. Whether conditions SW have been checked or not, algorithms should be designed to recognise the symptoms of *numerical* rank deficiency and, if detected, determine a *particular* least squares solution to the problem. It could be argued that in cases where conditions SW have been found to be violated, and hence R is not invertible, that it is unreasonable to demand a solution. We consider however that attention should be given to such cases for the simple reason that particular solutions may exist that provide perfectly acceptable spline approximants. Moreover, as noted, the mere satisfaction of conditions SW does not indicate the extent to which (8) is well-conditioned, which itself cannot be established until the system is formed. Therefore, since the formation of (8) normally constitutes the bulk of the work in obtaining a least squares solution, it seems perfectly reasonable to return a particular solution in rank deficient cases.

Resolving constraints for rank deficient problems

Rank deficient problems can be handled by appending to the observation equations or, equivalently, to the triangular system (8), appropriately selected equations known as *resolving constraints* (Gentleman, 1974). The number and form of the additional equations should be chosen such that the least squares solution of the augmented equations has the properties that

(i) it is unique and numerically well defined,

(ii) $e^T e$ is essentially unchanged.

In *exact* arithmetic, rank deficiency manifests itself as the appearance of *zero rows* in the triangular system (8). Suppose row j is zero for some j. Then, without further information, c_j cannot be determined. However, the system may be updated, using Givens rotations, for example, in no more than about qn operations, to incorporate the resolving constraint $g^T c = h$. Here g and h are respectively a vector and a scalar chosen (a) in accord with (i) and (ii) above, and (b) such that the updated system retains its band structure and has $r_{jj} \neq 0$. Moreover, since the value of r_{jj} is zero before the updating step, the updating can readily be shown to amount to the interchanging of the zero row with the constraint, the residual sum of squares $e^T e$ remaining unchanged.

For cases in which r_{jj} is nonzero but is judged to be "small", the system can still be updated in this way. However, rows j to n of the system will now normally be modified by the updating and $e^T e$ will accordingly increase slightly.

For each j for which the current r_{jj} is zero or "small", a resolving constraint is so introduced. The resulting triangular system will then be of full rank and can therefore be solved in the usual way. The form of solution obtained will be influenced by the choice of resolving constraints.

The reduced parameter solution

The reduced parameter solution results from the use of resolving constraints of the form $c_j = 0$, i.e. $g^T = (0, \ldots, 0, 1, 0, \ldots, 0)^T$, $h = 0$, with unity as the jth element of g. Such a solution will have some zero or "small" B-spline coefficients and thus essentially be equivalent to a conventional least squares solution obtained after rejecting certain elements from the B-spline basis that do not "contribute" to the solution. De Boor (1978) also proposes a strategy for the reduced parameter solution that is based upon the explicit formation of the normal equations (6) followed by a variant of Cholesky decomposition.

Shortcomings of the reduced parameter solution

We believe it important that spline solutions to practical data approximation problems should possess certain elementary invariance properties. The major shortcoming of the reduced parameter solution is that in rank deficient cases it is not invariant to certain simple transformations of the data. To illustrate this fact, consider the example discussed above that led to the matrix of form (9). It is readily verified that the resulting upper triangular matrix R has r_{jj}, $j = 1, \ldots, n$, of "reasonable" size and that $r_{n+1,n+1} = \delta^{n-1}$. Suppose that δ is sufficiently small for $r_{n+1,n+1}$ to be declared as zero. The resulting reduced parameter solution, leading to the spline $s_1(x)$ say, then has $c_{n+1} = 0$. However, if a constant K were added to all data ordinates (which corresponds merely to a different choice of origin for the dependent variable), and the problem re-solved, the B-spline coefficients of the spline $s_2(x)$ so produced would all be increased by K except for c_{n+1} which would remain zero. Thus, from the properties of B-splines, $s_2(x) \equiv s_1(x) + K$ for $x < \lambda_1$, but $s_1(x_{max}) = s_2(x_{max}) = 0$. $s_2(x)$ is not therefore simply a vertical translation of $s_1(x)$.

The above undesirable property can be overcome in this simple example by initially carrying out an appropriate vertical shift of the dependent variable values and subsequently modifying correspondingly the computed B-spline coefficients. In more complicated cases, however, this approach is inapplicable.

The minimum norm solution

A pàrticular solution that is frequently proposed in the context of *general* linear least squares problems is the *minimum norm solution*, which has the property that it minimizes c^Tc. We shall not give a description of algorithms for this solution here. It has however been used for an important class of spline approximation problems, viz. data fitting by bivariate splines of order 4. Hayes and Halliday (1974) discuss such problems using Householder transformations to effect solution. (The NAG Library (see Ford et al, 1979) includes an algorithm based on this paper but using an extension of the structure-exploiting Givens scheme referred to above (Hayes, 1978)).

The minimum norm solution suffers from similar shortcomings to those discussed for the reduced parameter solution. Indeed, as Hayes and Halliday indicate, the two solutions are identical when the rank deficiency is due solely to columns of elements which are all zero, or treated as zero as in the example above. Of course, these shortcomings relate solely to regions lacking data and it can be argued that a satisfactory approximation to the underlying function cannot reasonably be expected in such regions: the main aim is to ensure that the lack of data does not prevent a satisfactory solution in those regions where there is adequate data. However, in the following section, we propose an alternative with much better characteristics.

The reduced difference parameter solution

The reduced parameter solution can be viewed as a particular case of a solution we term the *reduced difference parameter solution*. Rather than impose constraints of the form $c_j = 0$ we consider the imposition of constraints $c_j^{(\ell)} = 0$ for an appropriate nonnegative value of ℓ. Here $c_j^{(\ell)}$ is related through I(23) to the values of c_j. We consider only *even* values of ℓ because this choice leads to an advantageous symmetry in the way in which constraints are imposed.

The case $\ell = 2$ is selected for the following reason. The choice $\ell = 0$ yields the reduced parameter solution already discussed. The choice $\ell \geq 4$ is applicable only to splines of order ≥ 5 (see I(23)). Apart from these considerations, the choice $\ell = 2$ proves to have some desirable properties, as will be discussed.

Recurrence I(23), with $\ell = 2$, then yields

$$\frac{c_{j+1} - c_j}{\lambda_j - \lambda_{j-n+1}} - \frac{c_j - c_{j-1}}{\lambda_{j-1} - \lambda_{j-n}} = 0,$$

a relation between three neighbouring B-spline coefficients, from which c_j can be expressed as a *convex combination* of c_{j-1} and c_{j+1}, viz.

$$c_j = (1-\mu)c_{j-1} + \mu c_{j+1}, \quad \mu = \frac{\lambda_{j-1} - \lambda_{j-n}}{\lambda_j - \lambda_{j-n+1} + \lambda_{j-1} - \lambda_{j-n}}. \tag{10}$$

Note that (10) is not defined if $\lambda_{j-n} = \ldots = \lambda_j$, in which case we arbitrarily use $c_j = \frac{1}{2}(c_{j-1} + c_{j+1})$ instead, the form to which (10) also reduces, automatically, in the case of equispaced knots. Also note that if $j = 1$ or $j = q$, a variant of (10) is used instead.

Properties of reduced difference parameter solutions

Valuable properties of the use of resolving constraints of the form $c_j^{(2)} = 0$ are as follows:

(i) When they are plotted at points equal to the arithmetic means of the interior knots of the B-splines associated with them, the B-spline coefficients tend to model the values of the function being approximated (de Boor, 1978). In cases where there is inadequate data in certain regions to define the B-spline coefficients that "relate" to those regions, it seems sensible therefore to determine such B-spline coefficients by choosing their values to be close in some sense to "neighbouring" coefficients that are better defined. The consequence of such a choice is a spline that will not be "pulled" towards zero, but one that has behaviour that is more consistent with that of a smooth function underlying the data.

(ii) Not only is the shape of the resulting spline invariant with respect to a vertical translation of the data but, if $\ell(x)$ denotes any straight line and $s_1(x)$ and $s_2(x)$ are respectively spline approximations to the data (x_i, f_i), $i = 1, \ldots, m$, and $(x_i, f_i + \ell(x_i))$, $i = 1, \ldots, m$, then $s_2(x) = s_1(x) + \ell(x)$.

(iii) Suppose $n - 2$ conditions $c_k^{(2)} = 0$, $k = j, \ldots, j + n - 3$, are introduced. From the compact support property of the B-splines it follows that $s''(x) \equiv 0$, $x \in (\lambda_{j-1}, \lambda_j)$, and hence that $s(x)$ is *linear* within this interval. Thus, in such a case, $s(x)$ is a polynomial spline of order n containing a straight line segment (cf Hayes, 1982). Such a spline would usually be more acceptable than one derived from the reduced parameter or minimum norm criteria.

A remark on multivariate spline approximation problems

The reduced difference parameter solution based upon resolving constraints of the form $c_j^{(2)} = 0$ serves as a model for the realistic treatment of rank deficiency in multivariate spline approximation problems. There is no obvious generalization of conditions SW to higher dimensions (except in special cases such as when the data lies at all meshpoints of a rectangular mesh) and hence, as pointed out by Hayes and Halliday (1974), it is essential to be able to treat rank deficiency in this case. Hayes and Halliday use the minimum norm solution for the bivariate case, but since this suffers from similar shortcomings to that in the univariate case, there would apparently be advantage to be gained from the use of reduced difference parameter solutions with resolving constraints analogous to the usual 5-point finite difference approximations to the Laplacian operator.

Choice of exterior knots

A discussion of the choice I(17), viz. $\lambda_j = x_{min}$, $j < 0$, and $\lambda_j = x_{max}$, $j > N + 1$, for the exterior knots is given in Cox (1975a) and Cox (1978b). The main conclusions, based upon a comparison with other choices, are as follows:

(i) The choice involves only the interval endpoints x_{min} and x_{max}, and can therefore be considered less arbitrary than choices that depend also upon the interior knots.

(ii) The spectral condition numbers of the matrices associated with a range of some twenty practical least squares spline approximation problems were found to be significantly smaller (Cox (1975b)). Moreover, Kozak (1980) has established theoretically that the condition number, with respect to the maximum norm, associated with the matrices occurring in spline interpolation problems is *minimal* for the choice I(17).

(iii) The imposition of boundary conditions in spline interpolation and approximation problems is greatly facilitated, as illustrated earlier.

Choice of knots for derivative-free spline interpolation

Any set of interior knots that satisfies conditions (2) will suffice, but the following choice is, in our experience, normally a good one compared with other choices in that (i) it typically gives a better-conditioned system (1), and (ii) the resulting spline interpolant is usually more satisfactory in its behaviour throughout $[x_1, x_m]$. If $n = 2k$, knots are located at points given by the rule

$$\lambda_j = x_{j+k}, \quad j = 1, \ldots, N,$$

and, if $n = 2k + 1$, at

$$\lambda_j = \tfrac{1}{2}(x_{j+k} + x_{j+k+1}), \quad j = 1, \ldots, N.$$

Variants are of course possible in particular circumstances. If, for example, $n = 4$ and it is required to reproduce a discontinuity in the first derivative at $x = x_r$, $3 < r < N - 2$, of the function underlying the data, a knot of multiplicity three may be located at this point in place of the simple knots at $x = x_j$, $j = r - 1, r, r + 1$.

A simple modification of the above rule has also proved successful in the case of interpolation with derivative boundary conditions.

Knot placement for least squares spline approximation

Several algorithms have been constructed for the automatic selection of the number and positions of the knots in least squares spline approximation problems.

They can be classified loosely into three classes: (i) *numerical*, (ii) *statistical* and (iii) *mathematical*. *Numerical* algorithms are those that regard the problem as one of *nonlinear least squares*, i.e. the residual

$$e_i = e_i(c, \lambda) = w_i(f_i - c_i N_{n,j}(\lambda, x_i)),$$

where $\lambda = (\lambda_1, \ldots, \lambda_N)^T$ is included as an additional argument of $N_{n,j}$ to emphasise the dependence of the latter upon the former, is regarded as a function of both c and λ. The problem

$$\min_{c,\lambda} e^T e \qquad (11)$$

is then "solved" using software for nonlinear least squares problems, possibly taking advantage of separability: $\min_{c,\lambda} e^T e \equiv \min_{\lambda} \min_{c} e^T e$. *Statistical* algorithms attempt to place the knots in such a way that as far as possible the resulting spline approximant reproduces the "signal" part of the data without following too closely its noise content. Such an approximant may be constructed by determining knots at locations which ensure that its residuals satisfy a statistical *trend test*. *Mathematical* algorithms are based upon theoretical results for the error of a spline approximant. It is possible to construct a spline approximant s(x) of order n to a function f(x) such that the error f(x) - s(x) depends upon *local* information, viz. the knot spacing and the value of $f^{(n)}$ in the neighbourhood of x.

Numerical algorithms

The following description is typical of numerical algorithms for automatic knot placement. The nonlinear least squares problem (11) is solved successively for N = 0, 1, ... until a member of the set of approximants so determined is deemed acceptable. In this process, the knots obtained for N = k - 1 may be used to provide initial estimates of the knots for N = k. For instance, the knots obtained for N = k - 1 could be taken, together with an additional value, chosen, for example, at a point in a region where the approximant makes the "biggest" contribution to the residual sum of squares.

The individual nonlinear least squares problems may be solved using an appropriate routine from a mathematical algorithms library. To ensure that knots selected by the routine permit B-splines to be defined it is necessary to impose inequality constraints I(3).

Although this approach can sometimes provide acceptable solutions, the following factors create difficulties. (i) Algorithms of this type must be used with discretion since they can be extremely expensive in terms of computer time, particularly if m and the final value of N are large. (ii) It is not obvious how to judge the acceptability of a member of the set of solutions provided by increasing values of N, unless the user is able to specify the noise level in his data, from which an acceptable residual sum of squares can be estimated. (iii) The nonlinear least squares routine may provide only a *local* minimum. It is the author's contention that there may be many local minima (Cox, 1971) and that this is the usual rather than the exceptional case. (iv) The solution provided may not lie within the space of acceptable approximants. For example, a spline s(x) may be required for which $s^{(r)}(x)$ is continuous for all r = 0, 1, ..., n - 2, whereas the computed solution may have multiple knots and thus possess too few continuous derivatives.

An example of a numerical algorithm is that presented by de Boor and Rice (1968). Their algorithm attempts to determine a cubic spline s(x) with as few knots as possible so that $e^T e$ is less than a prescribed positive number. They attack this problem in the above manner. Since the minimizing value of $e^T e$ either decreases strictly with increasing N or, for some N, is equal to zero (Rice, 1969), it follows that N can be increased until the above condition is satisfied.

Statistical algorithms

An example of a statistical algorithm is that due to Powell (1970) in which the interior knots are introduced adaptively according to a trend test. His approach involves the initial approximation of the data by a spline s(x) of order 4 with a small number of equispaced knots. The residuals $e_i = f_i - s(x_i)$ are then examined and if there are regions where a trend is indicated, i.e. the values of e_i are not distributed in a random manner about zero, further knots are inserted in the regions indicated by the test. A new spline s(x) upon the updated set of knots is then determined. The process is repeated until hopefully an acceptable approximation is obtained. A discussion of the behaviour of the algorithm on typical problems is given in Cox (1976). Briefly, the algorithm seems particularly suited to cases where there are several hundred data points and the underlying function has similar behaviour throughout the range. The algorithm's effectiveness to cope satisfactorily with more difficult sets of data is limited by a condition imposed for theoretical reasons upon the rate of change of adjacent subinterval lengths. It should be noted that Powell's algorithm is not based solely upon the least squares criterion, but contains additional smoothing terms to reduce the magnitudes of the discontinuities in the values of the third derivative at the interior knots. Compared with the algorithm of de Boor and Rice, Powell's algorithm is computationally inexpensive.

Mathematical algorithms

Mathematical algorithms are based upon the fact that, if the underlying function $f(x) \in C^n[x_{min}, x_{max}]$, there exists a spline s(x) of order n with specified interior knots $\lambda_1, \ldots, \lambda_N$ such that for $j = 0, \ldots, N$,

$$\max_{x \in I_j} |f(x) - s(x)| \leq K_n |P_j|^n \max_{x \in P_j} |f^{(n)}(x)|,$$

where K_n is a constant depending upon n, and P_j denotes the interval $[\lambda_{j-n+2}, \lambda_{j+n-1}]$ with $|P_j|$ its length (de Boor, 1978). An approximation to the spline s(x) that provides the minimum value of $\max|f(x)-s(x)|$, $x \in [x_{min}, x_{max}]$, is therefore given by selecting interior knots such that

$$|P_j|^n \max_{x \in P_j} |f^{(n)}(x)| = \text{constant}, \quad j = 0, \ldots, N.$$

Thus, to achieve such an approximation, the knots should be located such that the lengths of the subintervals P_j are inversely proportional to $\left\{\max_{x \in P_j} |f^{(n)}(x)|\right\}^{\frac{1}{n}}$. This problem cannot be solved exactly because $f^{(n)}(x)$ is unknown! However, de Boor (1974) gives an approximate method of solution that is iterative in nature. Each iteration involves estimating $f^{(n)}(x)$ from the spline s(x) constructed at the previous iteration, and using this information to provide new knot locations and hence a hopefully improved spline approximant. For some problems the approach can deliver a perfectly adequate approximant, but for others the results can be disappointing (see Cox (1976) for further details).

A new statistical algorithm

Finally, we describe in outline a new statistical algorithm for determining the number and locations of the knots. Full details will appear elsewhere. The only information that has to be supplied to the algorithm are the data points (x_i, f_i), $i = 1, \ldots, m$, and the required order n of the spline. For simplicity of presentation it is assumed here that all data points are weighted equally and are such that $x_1 < \ldots < x_m$. The algorithm consists of two phases.

The first phase determines a least squares piecewise polynomial approximant p(x) of order n. The number and locations of its breakpoints are determined such that p(x) has (i) residuals $f_i - p(x_i)$, $i = 1, \ldots, m$, that do not exhibit a statistical trend, and (ii) the smallest number of breakpoints consistent with (i). The data is processed from left to right. First, the greatest value of r_1 is determined such that the data (x_i, f_i), $i = 1, \ldots, r_1$, can be approximated by a polynomial of order n without a trend being indicated. Then the greatest value of r_2 is determined such that the data (x_i, f_i), $i = r_1 + 1, \ldots, r_2$, can be so approximated. The process is repeated until all data is exhausted, the resulting p(x) being composed of a set of polynomial pieces with no continuity, except by chance, between neighbouring pieces.

This complete process is iterated (typically 5-10 times) with respect to a parameter in the trend test, in such a way that the resulting function p(x) is identical to that which would have been obtained had the data been processed from right to left.

In the second phase the piecewise polynomial so produced is regarded as a spline approximant of order n with knots of multiplicity n at the breakpoints. Each of these multiple knots is replaced by a group of n simple knots nearby. A final least squares spline approximant s(x), of order n, with these knots, is then constructed in the usual way. Since the precise technique used in the knot replacement scheme is unlikely to be critical, we have used a fairly straightforward scheme having the property that conditions SW are automatically satisfied, with the result that s(x) is unique. Tests we have carried out so far with a pilot version of this algorithm have been most encouraging.

Summary

The second and final part of this paper has examined the use of B-splines for purposes of interpolation and least squares approximation by splines. Particular attention has been paid to the algorithmic aspects, with methods for handling rank deficient spline approximation problems and for automatic placement of the knots being discussed.

Acknowledgement

I am most grateful to J.G. Hayes for his comments on the paper.

References

BARRODALE, I. and YOUNG, A. 1967 A note on numerical procedures for approximation by spline functions. *Comput. J. 9*, 318-320.

BUTTERFIELD, K.R. 1976 The computation of all the derivatives of a B-spline basis. *J. Inst. Math. Appl. 17*, 15-25.

COX, M.G. 1971 Curve fitting with piecewise polynomials. *J. Inst. Math. Appl. 8*, 36-52.

COX, M.G. 1972 The numerical evaluation of B-splines. *J. Inst. Math. Appl. 10*, 134-149.

COX, M.G. 1975a *Numerical methods for the interpolation and approximation of data by spline functions*. London, City University, PhD Thesis.

COX, M.G. 1975b An algorithm for spline interpolation. *J. Inst. Math. Appl. 15*, 95-108.

COX, M.G. 1976 A survey of numerical methods for data and function approximation. *The State of the Art in Numerical Analysis*, D.A.H. Jacobs, Ed., London, Academic Press, 627-668.

COX, M.G. 1978a The representation of polynomials in terms of B-splines. *Proc. Seventh Manitoba Conference on Numerical Mathematics and Computing*, D. McCarthy and H.C. Williams, Eds., Winnepeg, University of Manitoba, 73-105.

COX, M.G. 1978b The incorporation of boundary conditions in spline approximation problems. *Lecture Notes in Mathematics 630: Numerical Analysis*, G.A. Watson, Ed., Berlin, Springer-Verlag, 51-63.

COX, M.G. 1978c The numerical evaluation of a spline from its B-spline representation. *J. Inst. Math. Appl. 21*, 135-143.

COX, M.G. 1981 The least squares solution of overdetermined linear equations having band or augmented band structure. *IMA J. Numer. Anal. 1*, 3-22.

CURRY, H.B. and SCHOENBERG, I.J. 1966 On Pólya frequency functions IV: the fundamental spline functions and their limits. *J. Analyse Math. 17*, 71-107.

DE BOOR, C. 1972 On calculating with B-splines. *J. Approx. Theory, 6*, 50-62.

DE BOOR, C. 1974 Good approximation by splines with variable knots II. *Lecture Notes in Mathematics 363: Numerical Solution of Differential Equations*, G.A. Watson, Ed., Berlin, Springer-Verlag, 12-20.

DE BOOR, C. 1976 Total positivity of the spline collocation matrix. *Indiana Univ. J. Math. 25*, 541-551.

DE BOOR, C. 1978 *A Practical Guide to Splines*. New York, Springer-Verlag.

DE BOOR, C. and PINKUS, A. 1977 Backward error analysis for totally positive linear systems. *Numer. Math. 27*, 485-490.

DE BOOR, C. and RICE, J.R. 1968 Least squares cubic spline approximation II - variable knots. Purdue University Report No. CSD TR 21.

FORD, B., BENTLEY, J., DU CROZ, J.J. and HAGUE, S.J. 1979 The NAG Library "machine". *Software-Practice and Experience 9*, 56-72.

GENTLEMAN, W.M. 1974 Basic procedures for large, sparse or weighted linear least squares problems. *Appl. Statist. 23*, 448-454.

HAYES, J.G. 1978 Data-fitting algorithms available, in preparation, and in prospect, for the NAG Library. *Numerical Software-Needs and Availability*, D.A.H. Jacobs, Ed., London, Academic Press.

HAYES, J.G. 1982 Curved knot lines and surfaces with ruled segments. To appear in proceedings of Dundee conference on Numerical Analysis, June 23-26, 1981, G.A. Watson, Ed., Berlin, Springer-Verlag.

HAYES, J.G. and HALLIDAY, J. 1974 The least-squares fitting of cubic spline surfaces to general data sets. *J. Inst. Math. Appl. 14,* 89-103.

KARLIN, S. 1968 *Total Positivity, Vol. 1.* Stanford, California, Stanford University Press.

KOZAK, J. 1980 On the choice of the exterior knots in the B-spline basis for a spline space. University of Wisconsin Report 2148.

POWELL, M.J.D. 1970 Curve fitting by splines in one variable. *Numerical Approximation to Functions and Data,* J.G. Hayes, Ed., London, Athlone Press, 65-83.

RICE, J.R. 1969 *The Approximation of Functions, Vol. II: Advanced Topics,* Reading, Mass., Addison-Wesley.

SCHOENBERG, I.J. and WHITNEY, Anne 1953 On Pólya frequency functions III. *Trans. Am. Math. Soc. 74,* 246-259.

FINITE ELEMENT METHODS FOR NON-SELF-ADJOINT PROBLEMS

K. W. Morton

1. INTRODUCTION

Finite element methods now dominate the solution of those elliptic problems that are derivable from quadratic extremal principles. Their practical development has been carried out by engineers under strong guidance from physical principles and subsequently their mathematical structure and error analysis studied in detail by mathematicians - as general references see Zienkiewicz (1977), Babuška & Aziz (1972), Strang & Fix (1973), Oden & Reddy (1976), Ciarlet (1978). From this work it became clear that the methods could be very widely applied, using a variational formulation referred to by the two groups respectively as 'the method of weighted residuals' or 'the weak form of the equations'.

However the success of these methods has sprung very largely from their optimal approximation properties. These follow naturally from extremal principles but are much harder to achieve for general elliptic problems that are not derived in this way and which are therefore no longer self-adjoint. It is this question that we shall consider in this short course of lectures. It is one of the most important and active areas of current research in the development of finite element methods.

1.1 A self-adjoint example

Consider the classical Dirichlet problem for Poisson's equation on an open polygonal domain Ω of $\mathbb{R}^2$ with boundary Γ :

$$-\nabla^2 u = f \quad \text{on } \Omega \qquad (1.1a)$$

$$u = 0 \quad \text{on } \Gamma. \qquad (1.1b)$$

Then u is also the solution of the extremal problem:

$$\underset{v \in H_0^1(\Omega)}{\text{minimise}} \int_\Omega [\tfrac{1}{2}|\underline{\nabla} v|^2 - fv]\,d\Omega \quad , \qquad (1.2)$$

where we denote by $H^m(\Omega)$ the usual Sobolev space of functions with square integrable m^{th} derivatives over Ω and by $H_0^m(\Omega)$ the closure in this space of the set of functions whose support is confined to the interior of Ω, i.e. which are zero on the boundary. (More generally we shall use the latter notation to denote functions which are zero on that part of the boundary where Dirichlet data is given).

Suppose now we triangulate Ω and denote by S_0^h the set of functions V which are continuous on Ω, linear in each triangle and zero on the boundary (in general that part with Dirichlet data): that is, we can write

$$V(\underline{x}) = \sum_{(j)} V_j \phi_j(\underline{x}), \tag{1.3}$$

where the summation is over all the interior vertices of the triangulation and $\{\phi_j\}$ are the basis functions which are pyramid-shaped: i.e., ϕ_j is piecewise linear, unity at node j and zero at all other nodes. Then the Ritz-Galerkin approximation U to u is given by

$$\underset{V \varepsilon S_0^h}{\text{minimise}} \int_\Omega [\tfrac{1}{2}|\underline{\nabla} V|^2 - fV] d\Omega \quad : \tag{1.4}$$

that is, using the notation $(\cdot,\cdot)$ for the L_2 inner product over Ω of either vectors or scalars, we have the Galerkin equations

$$(\underline{\nabla} U, \underline{\nabla}\phi_i) = (f, \phi_i) \qquad \forall \ \phi_i \ \varepsilon \ S_0^h. \tag{1.5}$$

Similarly, from (1.2) u satisfies the weak form of (1.1) :

$$(\underline{\nabla} u, \underline{\nabla} w) = (f,w) \qquad \forall \ w \ \varepsilon \ H_0^1(\Omega). \tag{1.6}$$

Since $S_0^h \subset H_0^1(\Omega)$, i.e. we are using a conforming finite element approximation, we can take ϕ_i for v in (1.6) and subtracting (1.5) obtain

$$(\underline{\nabla}(u-U), \underline{\nabla}\phi_i) = 0 \qquad \forall \ \phi_i \ \varepsilon \ S_0^h. \tag{1.7}$$

It follows, after a little manipulation, that

$$||\underline{\nabla}(u-U)||^2 = \inf_{V \varepsilon S_0^h} ||\underline{\nabla}(u-V)||^2 \ , \tag{1.8}$$

where $||\cdot||$ denotes the L_2 norm. This is the fundamental optimal approximation property of U.

In practical applications it is often the vector field $\underline{\nabla} u$ which is of most interest; and $\underline{\nabla} U$ is the least squares best fit to it from those piecewise constant approximations obtained by taking gradients of functions in S_0^h. By comparing with u^I, the piecewise linear interpolant of u, one readily finds that $\underline{\nabla} U$ is generally accurate to $O(h)$, where h is the maximal diameter of the triangles in the triangulation (and it is assumed that the latter satisfies a regularity condition such as all the angles are bounded from zero, or the weaker condition that they are bounded from π - see Strang & Fix, 1973, and Ciarlet, 1978.) On the other hand, from a well-known argument due to Aubin and Nitsche, for the error in U we have $||u-U|| = O(h^2)$. More generally, if S_0^h had contained all functions which were piecewise polynominal up to degree k in each triangle we would have $||u-U|| = O(h^{k+1})$ but $||\underline{\nabla}(u-U)|| = O(h^k)$.

However, a most important practical consideration is that superconvergence phenomena enable $\underline{\nabla}u$ to be estimated with an $O(h^2)$ error. For bilinear elements on rectangles $\underline{\nabla}U$ approximates $\underline{\nabla}u$ to $O(h^2)$ at the centroid of each element, a fact that has long been exploited by engineers and recently established rigorously and extended to more general quadrilateral elements by Zlamal (1977) - see also Zlamal (1978), Lesaint & Zlamal (1979). Although for triangular elements this extra order of convergence does not generally occur at the centroids, it has long been believed (and supported by numerical evidence) that second order accuracy can be recovered from gradients of U along each triangle side though this has not yet been proved (see Strang & Fix (1973) p169).

Without such superconvergence, which stems from the optimal approximation property, finite element methods would hardly be competitive with traditional finite difference methods, where similar divided difference results hold - see for instance Thomée & Westergren (1968).

1.2 Diffusion-convection problems

In studying the effects of losing self-adjointness we shall concentrate on the important class of problems called diffusion-convection problems:

$$-\underline{\nabla}.(a\underline{\nabla}u - \underline{b}u) + cu = f \qquad \text{in } \Omega \tag{1.9a}$$

$$u = g \quad \text{on } \Gamma_D \quad , \ \partial u/\partial n = 0 \qquad \text{on } \Gamma_N, \tag{1.9b}$$

where $\Gamma = \Gamma_D \cup \Gamma_N$, $\Gamma_D \cap \Gamma_N = \emptyset$ and $\Gamma_D \neq \emptyset$. Here the positive scalar a can be regarded as an isotropic diffusion coefficient, the vector $\underline{b}$ a convective velocity, f a given source and c a depletion rate. Equation (1.9a) represents a conservation law for the quantity u which might, for instance, be the concentration of a pollutant, the temperature of a coolant or density of a population: when $\underline{b} \neq \underline{0}$ it is not derivable from an extremum principle.

We may define an approximation $U(\underline{x})$ directly from Galerkin equations like (1.5), after first dealing with the inhomogeneous Dirichlet data on Γ_D. To ensure we maintain a strictly conforming approximation, we assume that the triangulation has been carried out so that Γ_D consists of a set of triangle sides with no more than one from any triangle: then define $G(\underline{x})$ as the piecewise blended interpolant which equals $g(\underline{x})$ on Γ_D, varies linearly in each of these triangles on the rays from Γ_D to the vertex not on Γ_D and is zero elsewhere. We define S_0^h, consistently with earlier usage, as the set of piecewise linear functions which are zero on Γ_D, and S_E^h as

$$S_E^h = \{V(\underline{x}) = G(\underline{x}) + W(\underline{x}) \mid W \varepsilon S_0^h\}. \tag{1.10}$$

Then the Galerkin approximation U to u is given by

$$(a\underline{\nabla}U,\underline{\nabla}\phi_i) + (\underline{\nabla}\cdot(\underline{b}U) + cU, \phi_i) = (f, \phi_i) \qquad \forall\, \phi_i \in S_0^h. \tag{1.11}$$

It is readily seen that u also satisfies these equations but, instead of an optimality property like (1.8), the most that we shall be able to deduce about the error is the following: defining the energy norm $||\cdot||_{AC}$ for the symmetric part of the operator by

$$||v||_{AC}^2 = (a\underline{\nabla}v, \underline{\nabla}v) + (cv, v), \tag{1.12}$$

we obtain for some constant K

$$||u-U||_{AC}^2 \le K \inf_{V\varepsilon S_E^h} ||u - V||_{AC}^2 . \tag{1.13}$$

The constant K will be bounded independently of h so that $\underline{\nabla}u$ will still have overall $O(h)$ accuracy, and indeed U will be $O(h^2)$ accurate, but the superconvergence phenomena are lost. Moreover, K depends on the local mesh Péclet numbers defined as bh/a, where b is the magnitude of $\underline{b}$, and these may be very large indeed. In practice the approximation may become very poor, exhibiting spurious oscillations which make it worthless.

1.3 A one-dimensional model problem

The origin of these oscillations can be exhibited by a simple model problem:

$$-au'' + bu' = 0 \qquad \text{on } (0,1) \tag{1.14a}$$

$$u(0) = 0, \quad u(1) = 1 \tag{1.14b}$$

where a,b are positive constants. The Galerkin equations (1.11) for a piecewise linear approximation on a uniform mesh of size h with $Jh = 1$, reduce to

$$h^{-1}(U_j-U_{j-1})(a+\tfrac{1}{2}bh) + h^{-1}(U_{j+1}-U_j)(-a+\tfrac{1}{2}bh) = 0 \tag{1.15a}$$

i.e. $\quad -\delta^2U_j + (bh/a)\,\Delta_0U_j = 0 \quad , \; j = 1,2..., J-1.$

The central differences here, $\delta^2U_j = U_{j+1}-2U_j+U_{j-1}$ and $\Delta_0U_j=\frac{1}{2}(U_{j+1}-U_{j-1})$, are typical of a Galerkin approximation and it has long been recognised that they may give rise to spurious oscillation when bh/a is large. In fact, we can solve this system explicitly to find for $j=0,1,...,J$ and $bh/a \neq 2$

$$U_j = \frac{\mu_0^j - 1}{\mu_0^J - 1} \qquad \text{where } \mu_o = \frac{2+bh/a}{2-bh/a} . \tag{1.16}$$

When $bh/a = 2$, we have $U_j = 0$ for $j = 0,1,...,J-1$ which is actually the exact solution of the reduced problem obtained by setting $a = 0$ in (1.14a): all the oscillatory solutions obtained from $bh/a > 2$ are entirely spurious and if one attempts to approximate the singular perturbation problem, $a \to 0$ with bh fixed, one finds that the Galerkin equations become singular when J is even and then $U_j \to \infty$ for j odd. The exact solution of (1.14) is

$$u(x) = \frac{e^{bx/a}-1}{e^{b/a}-1}, \quad 0 \le x \le 1, \tag{1.17}$$

giving an exponential boundary layer at $x \doteq 1$ as $a \to 0$, and (1.16) is a reasonable approximation only so long as μ_0 is a reasonable approximation to $e^{bh/a}$.

It might be argued, see for instance Gresho & Lee (1979), that it is unreasonable to attempt to approximate (1.14) when b/a is large without a local mesh refinement near the boundary layer. But in more complicated problems such layers are difficult to locate and the refinement expensive to implement. Thus most would agree that it is valuable to have available methods which will give good accuracy away from the boundary layer while using coarse meshes. What is certainly true is that, in various norms, the best piecewise linear fit is capable of giving an adequate representation of the solution under these circumstances: and with an appropriate choice of norm it can even give valuable information about the boundary layer, such as its half-width. Unfortunately, the Galerkin method does not give an approximation which is anywhere near optimal in any sense.

It should be noted in the above problem that, if the boundary condition at $x = 1$ were the Neumann condition, the exact solution would be identically zero. The Galerkin equations would have an extra equation $U_J - U_{J-1} = 0$, obtained from ϕ_J in (1.11), and their solution would also be identically zero for all bh/a. Thus the problem of spurious oscillations is a product of both the lack of self-adjointness and the boundary condition. We shall see however, in an example in Section 4, that even for a Neumann boundary condition the Galerkin method gives very poor accuracy compared with other methods.

1.4 Upwind differencing and Petrov-Galerkin methods

One-sided or upwind differencing has long been used in difference methods to avoid the oscillations described above. Completely replacing Δ_0 in (1.15b) by the backward difference operator Δ_- can however give rise to excessive false diffusion: writing $\Delta_- = \Delta_0 - \frac{1}{2}\delta^2$, the equations for U^- become

$$-(1+\tfrac{1}{2}bh/a)\,\delta^2 U_j^- + (bh/a)\,\Delta_0\, U_j^- = 0, \; j = 1,2,\dots,J-1 \tag{1.18}$$

with the solution of the same form as in (1.16) but with μ_0 replaced by $\mu_- = 1 + bh/a$. This is clearly always monotone but, for instance, for $bh/a = 2$ gives $U_{J-1}^- \approx 1/3$ rather than $u(1-h) \approx e^{-2}$, a typical example of the enhanced diffusion apparent from (1.18).

More sophisticated schemes, using exponential fitting, go back to Allen & Southwell (1955): here the technique gives

$$-(1+\tfrac{1}{2}\xi bh/a)\delta^2 U_j^e + (bh/a)\Delta_0 U_j^e = 0, \quad j=1,2,\ldots,J-1 \qquad (1.19)$$

where $$\xi = \coth(\tfrac{1}{2}bh/a) - (\tfrac{1}{2}bh/a)^{-1}, \qquad (1.20)$$

which exactly reproduces the nodal values $u(jh)$ of u. Moreoever, recent results have shown that by the use of local values of ξ in a variable coefficient problem one can obtain an approximation which is uniformly accurate at the nodes as $bh/a \to \infty$ - see Doolan et al. (1980) as a general reference for these developments.

Such results relate to one-dimensional problems. In higher dimensions difficulties occur with cross-wind diffusion, that is enhanced diffusion perpendicular to the velocity vector $\underline{b}$. Much less progress has been made here with finite difference methods.

A large part of the development of finite element methods for diffusion-convection problems has been inspired by the earlier work on difference methods. Several techniques for generating upwind schemes have been proposed and used quite successfully on two-dimensional problems. We shall consider these in more detail in Section 3. The earliest of them (Christie et al. 1976) is based on a generalisation of the Galerkin formulation in which a different set T_0^h of test functions ψ_i, is introduced in (1.11) instead of the trial function basis $\{\phi_i\}$: this Petrov-Galerkin approximation is then given by

$$(a\underline{\nabla}U, \underline{\nabla}\psi_i) + (\underline{\nabla}.(\underline{b}U) + cU, \psi_i) = (f, \psi_i) \qquad \forall\, \psi_i \in T_0^h. \qquad (1.21)$$

The problem is how to choose T_0^h for a given choice of S^h. Furthermore what criterion should be used for the assessment of accuracy and how should error bounds be derived? In particular, how closely should one adhere to the finite difference viewpoint, with the attendant emphasis on approximating nodal values and the awkwardness of estimating accuracy through estimating truncation error and bounding the inverse of the discrete operator?

All of the methods described in Section 3 to some extent adopt the finite difference viewpoint. In Section 4 an alternative approach is considered which is based on approximately symmetrising the bilinear form in (1.11). This leads to a near-optimal approximation to u in an integral norm which results naturally from the symmetrisation. We shall also show that the Section 3 methods can be regarded as approximate symmetrization in norm $||\cdot||_{AC}$. Thus the next section is devoted to developing the mathematical framework needed to study variational problems of the diffusion-convection type together with their approximation by generalised Galerkin procedures.

2. VARIATIONAL FORMULATION AND APPROXIMATION

2.1 Abstract problems and their approximation

The theoretical basis for Petrov-Galerkin methods is provided by the following generalisation of the Lax-Milgram lemma given by Babuška & Aziz (1972).

Theorem 2.1 Suppose $B(\cdot,\cdot)$ is a bilinear form on $H_1 \times H_2$, where H_1 and H_2 are real Hilbert spaces, which is continuous and coercive in the sense that there exist positive constants C_1 and C_2 such that

(i) $|B(v,w)| \leq C_1 ||v||_{H_1} ||u||_{H_2}$ $\quad \forall v \in H_1, \quad \forall w \in H_2;$ (2.1a)

(ii) $\inf_{v \in H_1} \sup_{w \in H_2} \frac{|B(v,w)|}{||v||_{H_1} ||w||_{H_2}} \geq C_2$; (2.1b)

(iii) $\sup_{v \in H_1} |B(v,w)| > 0 \quad \forall w \neq 0.$ (2.1c)

Then for $\forall f \in H_2'$, there is a unique $u_0 \in H_1$ such that

$$B(u_0,w) = f(w) \qquad \forall w \in H_2 \tag{2.2a}$$

and

$$||u_0||_{H_1} \leq ||f||_{H_2'}/C_2 \quad . \tag{2.2b}$$

Proof By (2.1a) and the Riesz representation theorem, for each $v \in H_1$ there is a Riesz representer Rv of $B(v,w)$ in H_2 such that

$$(Rv,w)_{H_2} = B(v,w) \qquad \forall v \in H_1, \ \forall w \in H_2 \tag{2.3a}$$

and also that

$$||R||_{\mathcal{L}(H_1,H_2)} \leq C_1 \ . \tag{2.3b}$$

That the mapping $R: H_1 \to H_2$ is closed follows from the closed graph theorem: furthermore, it follows from (2.1b) that

$$||Rv||_{H_2} = \sup_{w \in H_2} \frac{|B(v,w)|}{||w||_{H_2}} \geq C_2 \ ||v||_{H_1} \ . \tag{2.4}$$

Then by (2.1c) the mapping R must be onto: for otherwise, by the projection theorem, $\exists w^* \neq 0$ such that

$$(Rv,w^*)_{H_2} = 0 \qquad \forall v \in H_1$$

which contradicts (2.1c). From (2.4) we then have

$$||R^{-1}||_{\mathcal{L}(H_2,H_1)} \leq 1/C_2 \tag{2.5}$$

and if w_0 is the Riesz representer of f in H_2 we can write $u_0 = R^{-1} w_0$ to obtain (2.2a,b).∎

Corollary 1 If H_1 is a subspace of H_2 it is sufficient to replace (2.1b) by taking the supremum over H_1 and requiring that

$$|B(v,v)| \geq C_2'||v||^2_{H_1} \qquad \forall v\in H_1. \tag{2.6}$$

Corollary 2 If $H_1 = H_2$ and (2.6) is satisfied then Theorem 2.1 reduces to the Lax-Milgram lemma.

Theorem 2.2 (A generalisation of Céa's lemma). Suppose $B(\cdot,\cdot)$ on $H_1 \times H_2$, f and u_0 are as in Theorem 2.1 and that M_1, M_2 are subspaces of H_1, H_2 respectively such that, for some positive constant C_2^M :

(i) $$\inf_{V\in M_1} \sup_{W\in M_2} \frac{|B(V,W)|}{||V||_{H_1}||W||_{H_2}} \geq C_2^M ; \tag{2.7}$$

(ii) $$\sup_{V\in M_1} |B(V,W)| > 0 \qquad \forall W\neq 0,\ W\in M_2. \tag{2.8}$$

Then there is a unique $U_0\in\overline{M}_1$ given by

$$B(U_0,W) = f(W) \qquad \forall W\in M_2 \tag{2.9}$$

and moreover,

$$||u_0-U_0||_{H_1} \leq \left[1+C_1/C_2^M\right] \inf_{V\in M_1}||u_0-V||_{H_1} . \tag{2.10}$$

Proof With Rv defined as in Theorem 2.1, let P be the orthogonal projection $H_2\to\overline{M}_2$ and define S, in a similar way to R, as the mapping from $\overline{M}_1$ onto $\overline{M}_2$ such that

$$(SV,W)_{H_2} = B(V,W) \qquad \forall V\in\overline{M}_1,\ \forall W\in\overline{M}_2. \tag{2.11}$$

Then S is the restriction of PR to $\overline{M}_1$ because for $V\in\overline{M}_1$, $W\in\overline{M}_2$

$$(PRV,W)_{H_2} = (RV,W)_{H_2} = B(V,W).$$

Hence with w_0 the Riesz representer of f in H_2 and Pw_0 the representer in $\overline{M}_2$, we can set $U_0 = S^{-1}Pw_0 = S^{-1}PRu_0$ to obtain (2.9) from (2.11). Moreover, suppose V is any element of M_1, so that $S^{-1}PRV = V$, then we have

$$u_0 - U_0 = (I-S^{-1}PR)u_0 = (I-S^{-1}PR)(u_0-V);$$

$$\therefore \quad ||u_0-U_0||_{H_1} \leq ||I-S^{-1}PR|| \inf_{V\in M_1} ||u_0-V||_{H_1}$$

$$\leq [1+C_1/C_2^M] \inf_{V\in M_1} ||u_0-V||_{H_1} . \quad \blacksquare$$

Corollary 3 If H_1 is a subspace of H_2 and (2.6) holds and if $M_1=M_2$, then Theorem 2.2 reduces to Céa's lemma. For, from (2.2a) and (2.9), $B(u_0-U_0,V) = 0 \ \forall V\in M_1$: thus, by (2.1a) and (2.6),

$$||u_0-U_0||^2_{H_1} \le (C_2')^{-1}\ B(u_0-U_0,\ u_0-U_0) = (C_2')^{-1}B(u_0-U_0,u_0-V)$$

$$\le (C_1/C_2')||u_0-U_0||_{H_1}||u_0-V||_{H_1} \qquad \forall\ V\in M_1.$$

That is, (2.10) for this Galerkin case is replaced by

$$||u_0-U_0||_{H_1} \le (C_1/C_2') \inf_{V\in M_1} ||u_0-V||_{H_1}\ . \tag{2.12}$$

<u>Self-adjoint case</u>. If $B(\cdot,\cdot)$ is symmetric as well as coercive, it can be used to define an inner product and thence a Hilbert space H so that we set $H_1=H_2=H$. Then (2.1a) is replaced by the Cauchy-Schwarz inequality $|(v,w)_H| \le ||v||_H||w||_H$ with $C_1=1$ and (2.6) holds with $C_2'=1$. Thus, in Theorem 2.1, R becomes the identity mapping and the solution u_0 is just the Riesz representer of f in H. In the Theorem 2.2, S becomes the orthogonal projection of $\overline{M}_1$ onto $\overline{M}_2$ and $SU_0=Pu_0$. Moreoever, we can interpret C_2^M as measuring the extent to which elements of $\overline{M}_1$ can be approximated from $\overline{M}_2$: from (2.7) we have $C_2^M \le 1$ and, $\forall\ V\in M_1$,

$$||V||^2_H = ||V-SV||^2_H + ||SV||^2_H \ge ||V-SV||^2_H + (C_2^M)^2||V_H||^2$$

i.e. $$||V-SV||^2_H \le [1-(C_2^M)^2]||V||^2_H \qquad \forall\ V\in M_1. \tag{2.13}$$

If we denote by U_0^* the orthogonal projection of u_0 onto $\overline{M}_1$, i.e. the Galerkin approximation, we have

$$||u_0-U_0||^2_H = ||u_0-U_0^*||^2_H + ||U_0^*-U_0||^2_H. \tag{2.14}$$

Then, rewriting the last term, recalling that $(u_0-U_0,W)_H = 0\ \forall\ W\in M_2$ and using (2.13), we obtain

$$||U_0^*-U_0||^2_H = (u_0-U_0,U_0^*-U_0)_H$$

$$= (u_0-U_0,(I-S)(U_0^*-U_0))_H$$

$$\le ||u_0-U_0||_H[1-(C_2^M)\ ]^{\frac{1}{2}}||U_0^*-U_0||_H$$

i.e. $$||U_0^*-U_0||_H \le [1-(C_2^M)^2]^{\frac{1}{2}}||u_0-U_0||_H. \tag{2.15}$$

Hence we obtain the error bound for the Petrov-Galerkin method in this case,

$$||u_0-U_0||_H \le (1/C_2^M)\ ||u_0-U_0^*||_H. \tag{2.16}$$

This is sharper than the bound obtained by merely putting $C_1 = 1$ in (2.10). In particular, as $C_2^M \to 1$ to give the Galerkin case, this error constant correctly tends to unity.

It was the fact that R became the identity which enabled this argument to go through so simply. In the general case, we could introduce the adjoint operator R^* to R and consider approximating $V \in M_1$ from R^*M_2 : if $W_V \in \bar{M}_2$ gives the best approximation,

$$||V-R^*W_V||^2_{H_1} = ||V||^2_{H_1} - ||R^*W_V||^2_{H_1}$$

and

$$||R^*W_V||_{H_1} = \sup_{W \in M_2} \frac{(R^*W_V, R^*W)_{H_1}}{||R^*W||_{H_1}} = \sup_{W \in M_2} \frac{(V, R^*W)_{H_1}}{||R^*W||_{H_1}}$$

$$\geq \frac{B(V,SV)}{||R^*SV||_{H_1}} \geq \frac{||SV||^2_{H_2}}{C_1||SV||_{H_2}} \geq \frac{C_2^M}{C_1} ||V||_{H_1}$$

i.e.
$$||V-R^*W_V||^2_{H_1} \leq [1-(C_2^M/C_1)^2]||V||^2_{H_1} \qquad \forall\ V \in M_1. \tag{2.17}$$

Also from $B(u_0-U_0, W) = 0, \quad \forall W \in M_2$, we have

$$(u_0-U_0, R^*W)_{H_1} = 0 \qquad \forall\ W \in M_2$$

so that if U_0^* is the orthogonal projection of u_0 onto $\bar{M}_1$ (but not now the Galerkin approximation),

$$||U_0^*-U_0||^2_{H_1} = (u_0-U_0, U_0^* - U_0)_{H_1}$$

$$= \inf_{W \in M_2} (u_0-U_0, U_0^*-U_0-R^*W)_{H_1}$$

$$\leq ||u_0-U_0||_{H_1} [1-(C_2^M/C_1)^2]^{\frac{1}{2}} ||U_0^*-U_0||_{H_1} .$$

Thus in the same way as (2.16) we obtain the following corrolorary: the derivation also indicates more clearly than (2.7) the appropriate choice of M_2 to ensure that U_0 is a near optional approximation to u_0 from M_1. We will use this later.

<u>Corollary 4</u> In Theorem 2.2 the error bound (2.10) can be improved to

$$||u_0-U_0||_{H_1} \leq (C_1/C_2^M) \inf_{V \in M_1} ||u_0-V||_{H_1} . \tag{2.18}$$

2.2 Diffusion-convection problems

We recall the general statement of the problem in (1.9a,b) and shall henceforth assume that Neumann boundary conditions are never imposed on inflow boundaries: that is, if $\underline{n}$ is the unit outward normal to Γ, then

$$\underline{n}.\underline{b} \geq 0 \qquad \text{on} \quad \Gamma_N. \tag{2.19}$$

It is also useful to denote by Γ^- the inflow boundary, i.e. that part of Γ on which $\underline{n}.\underline{b} < 0$, and similarly by Γ^+,Γ^0 the outflow and tangential boundaries. So we have assumed that $\Gamma_N \cap \Gamma^- = \emptyset$.

In making such an assumption we have also assumed some smoothness of the boundary. It will be sufficient to assume throughout that Γ is Lipschitz continuous in the sense of Nečas (1967) - see also Ciarlet (1978) and Oden & Reddy (1976) as general references for results needed here: that is, there are a finite number of local co-ordinate systems such that every part of the boundary is defined by a Lipschitz continuous function in at least one of them. This means that $\underline{n}$ is defined almost everywhere on Γ, which may have corners and edges but no cusps. It also means that a trace operator tr is defined on $H^1(\Omega)$, extending the restriction of $v:\bar{\Omega} \to \mathbb{R}$ to Γ as a continuous linear mapping $\text{tr} : H^1(\Omega) \to L_2(\Gamma)$. Thus we can write in a conventional way

$$H_0^1(\Omega) = \{v \in H^1(\Omega) | \text{tr } v = 0 \text{ on } \Gamma_D\}. \qquad (2.20)$$

By implication, too, Ω is bounded and a Poincaré-Friedrichs inequality holds: there exists a positive constant $C(\Omega)$ such that

$$||v||_{0,\Omega} \le C(\Omega)|v|_{1,\Omega} \qquad v \in H_0^1(\Omega), \qquad (2.21)$$

where the notation $||\cdot||_{m,\Omega}$ is used for the usual Sobolev norm for $H^m(\Omega)$ and $|\cdot|_{i,\Omega}$ for $i=1,\ldots,m$ denotes the corresponding semi-norms. Finally, Green's formulae also hold: for example, with $u \in H^1(\Omega)$, $\underline{v} \in [H^1(\Omega)]^d$, $\Omega \subset \mathbb{R}^d$, d=1,2 or 3 we have

$$\int_\Omega u \underline{\nabla} \cdot \underline{v} d\Omega = -\int_\Omega \underline{v} \cdot \underline{\nabla} u d\Omega + \int_\Gamma u \underline{n} \cdot \underline{v} d\Gamma. \qquad (2.22)$$

To apply (2.22) to (1.9) we assume that

$$0 < a \in C^0(\bar{\Omega}), \ \underline{b} \in [H^1(\Omega)]^d \text{ and } 0 \le c \in L_2(\Omega). \qquad (2.23)$$

Then, denoting by L the diffusion-convection operator on the left of (1.9a), we have for $v \in H^2(\Omega)$, $w \in H^1(\Omega)$:

$$(Lv,w) + (a\underline{n}\cdot\underline{\nabla}v,w)_{\Gamma_N} = (a\underline{\nabla}v-\underline{b}v, \underline{\nabla}w) + (cv,w) \\ +(\underline{n}\cdot(\underline{b}v-a\underline{\nabla}v),w)_{\Gamma_D} + (\underline{n}\cdot\underline{b}v,w)_{\Gamma_N}, \qquad (2.24)$$

where $(\cdot,\cdot)_{\Gamma_N}$ denotes the L_2 inner product over Γ_N and similarly for Γ_D. A more convenient basic definition of the bilinear form to which we shall apply Theorems 2.1 and 2.2 than (2.24) is however the following:

$$B(v,w) := (a\underline{\nabla}v,\underline{\nabla}w) + (\underline{\nabla}\cdot(\underline{b}v) + cv,w) - (a\underline{n}\cdot\underline{\nabla}v,w)_{\Gamma_D}. \qquad (2.25)$$

This is clearly continuous on $H^1(\Omega) \times H_0^1(\Omega)$ which covers most cases of interest: in Section 4, however, we shall try lifting the condition that $w=0$ on $\Gamma_D \cap \Gamma^-$

and this will entail restricting the class of v's.

The inhomogeneous boundary data of (1.9) is incorporated in the formulation by assuming that g is derived by the trace operator $(tr)_D$ corresponding to Γ_D operating on some function $G:\Omega\to\mathbb{R}$. Then we define the following linear functional corresponding to all the data:

$$F(w) := (f-\underline{\nabla}\cdot(\underline{b}G) - cG, w) - (a\underline{\nabla}G, \underline{\nabla}w) + (a\underline{n}\cdot\underline{\nabla}G, w)_{\Gamma_D} . \tag{2.26}$$

To ensure that this is bounded over $w\in H_0^1(\Omega)$ it is sufficient to assume:

$$f\in L_2(\Omega) , \quad g=(tr)_D G \quad \text{s.t.} \quad G\in H^1(\Omega). \tag{2.27}$$

Again, widening the class of w will entail restricting that of G. We shall furthermore assume that the convective medium is incompressible in obtaining the following Theorem.

Theorem 2.3 Suppose that for the problem (1.9a,b) the assumptions (2.19), (2.23) and (2.27) are satisfied and that

$$\underline{\nabla}\cdot\underline{b} = 0. \tag{2.28}$$

Then a weak solution exists of the form $u=u_0+G$, where $u_0\in H_0^1(\Omega)$ is uniquely defined by

$$B(u_0,w) = F(w) \qquad \forall w\in H_0^1(\Omega) \tag{2.29}$$

and $B(\cdot,\cdot)$, $F(\cdot)$ are defined by (2.22) and (2.24).

Proof We check the hypotheses of Theorem 2.1, or rather of the Lax-Milgram lemma, with $H_1=H_2=H_0^1(\Omega)$. For the coercivity we have

$$B(v,v) = (a\underline{\nabla}v,\underline{\nabla}v) + (cv,v) + (\underline{\nabla}\cdot(\underline{b}v),v) ;$$

and from (2.28), (2.19) there follows for $u\in H_0^1(\Omega)$

$$(\underline{b}\cdot\underline{\nabla}v,v) = (\underline{\nabla}\cdot(\underline{b}v),v) = -(\underline{b}\cdot\underline{\nabla}v,v) + (\underline{n}\cdot\underline{b}v,v)_{\Gamma_N}$$

i.e. $$(\nabla\cdot(\underline{b}v),v) = \tfrac{1}{2}(\underline{n}\cdot\underline{b}v,v)_{\Gamma_N} \geq 0. \tag{2.30}$$

Since we have $a>0$, $c\geq 0$ and $\Gamma_D \neq \emptyset$, (2.21) ensures that (2.6) is satisfied for some C_2'. Similarly, and with the use of the Cauchy-Schwarz inequality, the continuity condition (2.1a) is satisfied for some C_1. We defer until the next sub-section discussion on the sharpest bounds attainable for the ratio C_1/C_2 except to note that this will depend on bounding $(\underline{b}\cdot\underline{\nabla}v,w)$ in terms of $(a\underline{\nabla}v,\underline{\nabla}v)$ and $(a\underline{\nabla}w,\underline{\nabla}w)$.

As this solution u is clearly uniquely defined, independently of the choice of G, and since a classical solution of (1.9a,b) because of (2.24) also satisfies (2.29), identification of the weak and classical solutions depends only on the regularity of the latter. For general theorems covering this we refer the reader to Agmon, Douglis & Nirenburg (1964).

2.3 Galerkin approximation

The analysis of finite element methods takes its simplest form if we make the following standard assumptions regarding the approximation properties of the space S^h of trial functions:

(i) S^h is conforming, i.e. $S^h \subset H^1(\Omega)$;

(ii) S^h has order $r > 1$ and is regular, i.e. $\forall g \in H^\ell(\Omega), \ell \geq 1$, there exists an element $V \in S^h$ such that for some constant K

$$||g-V||_{s,\Omega} \leq Kh^\mu ||g||_{\ell,\Omega} \qquad s = 0,1 \qquad (2.31)$$

where $\mu = \min(r-s, \ell-s)$.

To ensure these properties we shall assume Γ is polygonal and Ω is subdivided into elements which have maximal diameter h and satisfy a regularity condition, such as all interior angles are uniformly bounded from zero. In $\mathbb{R}^2$, elements will either be triangles or quadrilaterals, with corresponding elements in $\mathbb{R}^3$: parametric transformations will generally be necessary to map quadrilaterals in global variables into rectangles in local variables and such transformations (c.f. isoparametric elements) are often used to approximate curved boundaries, but this is beyond the scope of the present lectures.

We shall mainly consider piecewise linear approximation on triangles or bilinear approximation on rectangles, both of which are examples of $r=2$ in (2.31). Piecewise quadratic, or biquadratic, elements similarly give $r=3$: generally speaking, if all polynomial functions up to degree k on each element are contained in S^h, and the parameters are chosen to ensure continuity between elements, then (2.31) will hold with $r=k+1$. We also omit consideration of Hermitian elements, such as Hermite cubics, so ensuring that we can write the general member $V \in S^h$ in the Lagrangian form

$$V(\underline{x}) = \sum_{(j)} V_j \phi_j(\underline{x}), \qquad (2.32)$$

where ϕ_j is the basis function corresponding to node $(\underline{x}_j)$ such that $\phi_j(\underline{x}^i) = \delta_{ij}$ and hence $V_j = V(\underline{x}_j)$. Thus $S^h = \text{span}\{\phi_j\}$.

We suppose further that Γ_D is composed of an integral number of element sides so that, defining

$$S_0^h = S^h \cap H_0^1(\Omega), \qquad (2.33)$$

we find that S_0^h is spanned by a subset of $\{\phi_j\}$. Then we can introduce the Galerkin approximation U_0 to the solution u_0 of (2.29),

$$U_0 \in S_0^h : \ B(U_0, \phi_i) = F(\phi_i) \qquad \forall \phi_i \in S_0^h . \qquad (2.34)$$

Thus we have

$$B(u_0 - U_0, \phi_i) = 0 \qquad \forall \phi_i \in S_0^h . \qquad (2.35)$$

Let us estimate the error u_0-U_0 first in the norm $||\cdot||_{AC}$ introduced in (1.12). From the assumptions (2.23) on a, $\underline{b}$ and c and the boundedness of Ω we can introduce constants P_1, P_2 such that

$$|(\underline{\nabla}\cdot(\underline{b}v),w)| \le P_1||v||_{AC}||w||_{0,\Omega} \tag{2.36a}$$

and

$$||w||_{0,\Omega} \le P_2||w||_{AC}. \tag{2.36b}$$

We can regard the product P_1P_2 as a global Péclet number, particularly when c=0, as it has the dimension of bL/a where L is a scale length. We also suppose that $u_0 \in H^r(\Omega)$ and that by (2.31) there therefore exists a member $\tilde{U}_0$ of S_0^h such that for some constants K and P_3

$$||u_0-\tilde{U}_0||_{0,\Omega} \le K_r h^r ||u_0||_{r,\Omega} \tag{2.37a}$$

and

$$||u_0-\tilde{U}_0||_{AC} \le K_r P_3 h^{r-1} ||u_0||_{r,\Omega} . \tag{2.37b}$$

Then from (2.30), (2.35) and (2.36a)

$$\begin{aligned} ||u_0-U_0||^2_{AC} &\le B(u_0-U_0,u_0-U_0) = B(u_0-U_0,u_0-\tilde{U}_0) \\ &= (u_0-U_0,u_0-\tilde{U}_0)_{AC} + (\underline{\nabla}\cdot[\underline{b}(u_0-U_0)],u_0-\tilde{U}_0) \\ &\le ||u_0-U_0||_{AC}\left[||u_0-\tilde{U}_0||_{AC}+P_1||u_0-\tilde{U}_0||_{0,\Omega}\right]; \end{aligned} \tag{2.38}$$

that is, by (2.37)

$$||u_0-U_0||_{AC} \le [1+P_1h/P_3]\, K_r P_3 h^{r-1} ||u_0||_{r,\Omega} . \tag{2.39}$$

Here we can regard P_1h/P_3 as a mesh Péclet number which is seen to completely represent the loss of accuracy attributable to the convection term when the Galerkin method is used. To compare this result with that obtained directly from Céa's lemma, (2.12), we see from (2.36) that C_1 in (2.1a) can be taken as $1+P_1P_2$ and C_2' from (2.27) taken as unity: thus we have been able to replace a global Péclet number with a mesh Péclet number.

To obtain an error estimate in a lower order norm, in particular in the L_2 norm, we use the device due to Aubin and Nitsche. We let $v_0 \in H^2(\Omega)$ be the solution of the adjoint problem

$$B^*(v_0,w) = (u_0-U_0,w) \qquad \forall w \in H_0^1 \tag{2.40}$$

for which, by the ellipticity of the equation, there must be an estimate of the form

$$||v_0||_{2,\Omega} \le P_4||u_0-U_0||_{0,\Omega} . \tag{2.41}$$

Then taking $w=u_0-U_0$ in (2.40) we obtain

$$||u_0-U_0||^2_{0,\Omega} = B^*(v_0,u_0-U_0) = B(u_0-U_0,v_0)$$
$$= B(u_0-U_0,\ v_0-\tilde{V}_0)$$
$$\leq ||u_0-U_0||_{AC}\Big[||v_0-\tilde{V}_0||_{AC} + P_1||v_0-\tilde{V}_0||_{0,\Omega}\Big]. \qquad (2.42)$$

Here, $\tilde{V}_0$ is any member of S^h_0 and we can assume it is chosen so that bounds analogous to (2.37) hold with $r=2$. Then substituting also from (2.39) and (2.41) we obtain

$$||u_0-U_0||_{0,\Omega} \leq [1+P_1h/P_3]^2 \quad K_rK_2P_3P_4h^r||u_0||_{r,\Omega}\ . \qquad (2.43)$$

Thus the usual extra power of h is obtained but at the cost of extra constants, in particular a further factor from the mesh Péclet number. Note that as $a\to 0$, although $P_4 \sim a^{-1}$ we also have $P_3 \sim a^{\frac{1}{2}}$ and so $P_3{}^2P_4 = O(1)$.

It is worth noting that this same technique can be used to obtain the pair (2.37 a and b). Suppose U^*_0 is the optimal approximation to u_0 in $||\cdot||_{AC}$, that is

$$(u_0-U^*_0,\phi_i)_{AC} = 0 \qquad\qquad \forall\ \phi_i\in S^h_0. \qquad (2.44)$$

Then we can introduce w_0 by

$$(w_0,w)_{AC} = (u_0-U^*_0,w) \qquad\qquad \forall\ w\in H^1_0 \qquad (2.45)$$

for which we shall have a bound $||w_0||_{2,\Omega} \leq P^*_4||u_0-U^*_0||_{0,\Omega}$ and an optimal approximation W^*_0 with $||w_0-W^*_0||_{AC} \leq K_2P^*_3\ h||w_0||_{2,\Omega}$. Hence we obtain, in the same way as (2.42),

$$||u_0-U^*_0||_{0,\Omega} \leq K_2P^*_3P^*_4h||u_0-U^*_0||_{AC} \qquad (2.46)$$

and substituting this in (2.38) obtain

$$||u_0-U_0||_{AC} \leq [1+K_2P_1P^*_3P^*_4h]||u_0-U^*_0||_{AC}\quad . \qquad (2.47)$$

We again see that $K_2P_1P^*_3P^*_4h$ can be regarded as a mesh Péclet number and, with a bound on $||u_0-U^*_0||_{AC}$, (2.47) can be used to replace (2.39).

Thus, too (2.47) shows that as $h\to 0$ the Galerkin approximation U_0 eventually becomes "near optimal" and one can expect superconvergence results to hold: the practical difficulty is that this will occur for only extremely small h when the Péclet number is large.

2.4 The one-dimensional model problem

We conclude this section by applying some of the results in the earlier subsections to the model problem (1.14). We reformulate and generalise this to

$$-au_0'' + bu_0' = f, \qquad\qquad u_0(0) = u_0(1) = 0, \qquad (2.48)$$

with $f\equiv -b$ giving for $u=u_0+x$ the same result as (1.14). Working in $H_0^1(0,1)$ equipped with the norm $||v||^2_{AC} = a||v'||^2_0$, the mapping R of Thorem 2.1 can be explicitly derived:

$$B(v,w) = \int_0^1 (av'w' + bv'w)dx = \int_0^1 a(Rv)'\, w'dx \qquad \forall\, v,w\in H_0^1$$

i.e. $$a(Rv)' - av' + bv = \text{const.} = b\overline{v}$$

i.e. $$(Rv)(x) = v(x) - (b/a)\int_0^x [v(t)-\overline{v}]dt, \tag{2.49}$$

where $\overline{v} = \int_0^1 vdt$: R^* has the same form with the sign of b changed. Similarly we find

$$(R^{-1}w)(x) = \int_0^x e^{b(x-t)/a}\, w'(t)dt - [1-e^{-b/a}]^{-1}\,[e^{bx/a}-1]\int_0^1 e^{-bt/a} w'(t)dt. \tag{2.50}$$

It is clear directly from (2.6) that $C_2'=1$ and from (2.49) that

$$||Rv||^2_{AC} = ||v||^2_{AC} + (b^2/a)\int_0^1 (v-\overline{v})^2 dx \quad : \tag{2.51}$$

it is therefore evident that $||R^{-1}||=1$ and from a Fourier analysis one can show that

$$||R|| = (1+ b^2/4\pi^2 a^2)^{\frac{1}{2}}. \tag{2.52}$$

This then is the constant which appears in Céa's lemma (2.12).

For the Galerkin approximation using piecewise linear elements on a uniform mesh we can also carry out the analysis leading to (2.47). It is easy to see that $P_1=b/a^{\frac{1}{2}}$ and $P_4^*=1/a$ and elementary approximation theory gives $K_2=1/\pi$ with $P_3^*=a^{\frac{1}{2}}$. Thus (2.47) becomes

$$||u_0-U_0||_{AC} \le [1 + bh/a\pi]\; ||u_0-U_0^*||_{AC}, \tag{2.53}$$

a much sharper result than that given by Céa's lemma.

Moreover, u_0-U_0 is given explicitly by (1.16) and (1.17), and it is easy to see (cf. (3.1) below) that U_0^* actually interpolates u_0. Thus we can readily calculate the ratio of the two norms in (2.53): we find that

$$||u_0-U_0||^2_{AC} = b(1-\mu_0^{-J})^{-1}\frac{\mu-\mu_0}{1-\mu\mu_0} + O(e^{-b/a}), \tag{2.54a}$$

$$||u_0-U_0^*||^2_{AC} = b\left[\frac{1}{2} - \frac{a}{bh}\;\frac{\mu-1}{\mu+1}\right] + O(e^{-b/a}), \tag{2.54b}$$

where μ_0 is given by (1.16) and $\mu=e^{bh/a}$, so that the ratio does not take a simple form. However, denoting bh/a by β, the two limiting forms are as follows:

$$||u_0-U_0||_{AC}\big/||u_0-U_0^*||_{AC} \sim 1+\frac{23}{240}\beta^2 \qquad \text{as} \quad \beta\to 0 \tag{2.55a}$$

$$\sim (\tfrac{1}{2}h\beta)^{\frac{1}{2}} \qquad \text{as} \quad \beta\to\infty,\ \text{even } J. \tag{2.55b}$$

Apart from the fact that it diverges, this second limit is not particularly useful since even U_0^* is a very poor approximation to u_0 in this norm: indeed this limit follows directly from the observations that $||U_0||_{AC}/||u_0||_{AC} \to (\frac{1}{2}h\beta)^{\frac{1}{2}}$ and $||U_0^*||_{AC}/||u_0||_{AC} \to (2/\beta)^{\frac{1}{2}}$ as $\beta \to \infty$.

A more dramatic demonstration of the inadequacy of the Galerkin approximation when β is large is provided by looking at the discrete equations which represent the approximation process $U_0 = S^{-1}Pw_0$. Here w_0, the Riesz representer of $-b$ in H_{AC}, is given by

$$w_0(x) = -\tfrac{1}{2}(b/a)x(1-x)$$

and P corresponds to taking nodal values. Then from equation (2.49) for R, we obtain for $SU_0 \equiv PRU_0 = Pw_0$

$$U_j^0 - \beta\left[\sum_1^{j}{}' U_i^0 - jh\sum_1^J U_i^0\right] = -\tfrac{1}{2}\beta hj(J-j),\ j=1,2,\ldots,J-1; \tag{2.56}$$

here $\{U_j^0,\ j=0,1,\ldots,J\}$ are the nodal values of $U_0(x)$ and the prime on the first sum indicates that only $\frac{1}{2}U_j^0$ is included. It is readily seen that for even J the vector $\{0,1,0,1,\ldots,1,0\}$ is annihilated by the operations in the square brackets. This is the vector for which the norm $||S^{-1}||=1$ is attained and there will always be a component of this of order of magnitude β in the solution U_0: thus it is that for even J the Galerkin solution exhibits unbounded oscillations as $\beta \to \infty$. One can similarly see why for odd J the oscillations are much less violent.

3. PETROV-GALERKIN METHODS USING EXPONENTIAL, UPWINDING AND STREAMLINE-DIFFUSION TECHNIQUES

In surveying these three (overlapping) techniques, we shall generally introduce them for the 1D model problem (1.14) before indicating their developments for more general problems. We shall also work in the space H_{AC} with norm $||\cdot||_{AC}$ defined in (1.12). This is a particularly appropriate norm in this case, in view of the motivation of several of the key ideas by finite difference methods. For, if the trial space S_E^h consists of piecewise linear functions with nodes $\{x_j\}$, $j = 0,1,\ldots,J$, $x_0=0$, $x_J=1$, the best fit $U^* \in S_E^h$ to u in this norm satisfies

$$a\int_0^1 (u' - U^{*\prime})\phi_j'\,dx = 0,\ j = 1,2,\ldots,J-1$$

i.e.
$$\frac{\Delta_-[u(x_j)-U_j^*]}{\Delta_- x_j} = \frac{\Delta_+[u(x_j)-U_j^*]}{\Delta_+ x_j}\ . \tag{3.1}$$

Denoting the common value by D, we find by multiplying each ratio by the denominator and summing that $D=0$: hence $u(x_j)-U_j^* = \text{constant} = 0$. Thus <u>the best piecewise linear fit in this norm is also the best (i.e. exact) fit at the nodes</u>.

3.1 Use of piecewise exponentials

As with finite differences, several early methods exploited the exponential character of solutions to one dimensional diffusion-convection problems. Three or four differing approaches have been adopted.

(i) Liouville transform (Guymon, 1970). If in the 1D model problem (1.14) we set

$$w(x) = e^{-\frac{1}{2}bx/a}u(x) \tag{3.2a}$$

the problem is symmetrized to

$$-aw'' + (b^2/4a)w = 0, \quad w(0) = 0, \quad w(1) = e^{-\frac{1}{2}b/a} . \tag{3.2b}$$

This may then be solved by a Galerkin method to give a best fit in the mixed norm

$$\int_0^1 [aw'^2 + (b^2/4a)w^2]dx, \tag{3.3a}$$

which tends to the L_2 best fit as $b/a \to \infty$. However, any errors will be amplified by $\exp(\frac{1}{2}bx/a)$ on transforming back to the original variables and this is ill-conditioned in the singular perturbation limit. Equivalently, we can see that after transforming back we have an optimal approximation in the norm

$$\int_0^1 av'^2 \, e^{-bx/a}dx \tag{3.3b}$$

which concentrates attention away from the boundary layer near $x=1$ which is of most interest. Guymon et al. (1970) have also extended this technique to two-dimensional flow problems, but the above arguments indicate that it should be used only with very great care.

(ii) Exponential trial space (K.E. Barrett, 1974, 1977). When any inhomogenous term in the equation is such that the solution is predominantly exponential in character, the following basis functions (on a uniform mesh) would seem a natural choice for the trial space: with $\beta=bh/a$ and $\phi_j(x)=\phi(h^{-1}x-j)$ we set

$$(1-e^{-\beta})\phi(t) = \begin{cases} e^{\beta t} - e^{-\beta} & -1 \le t \le 0 \\ 1 - e^{\beta(t-1)} & 0 \le t \le 1. \end{cases} \tag{3.4}$$

The Galerkin method for the model problem then of course gives the exact exponential solution. For the more general problem $-au''+ bu' = f(x)$, $u(0)$ and $u(1)$ given, then by (2.12) and (2.52) the approximation U has an error bound

$$||u-U||_{AC} \le (1 + b^2/4\pi^2a^2)^{\frac{1}{2}} \inf_{V\in S_E^h} ||u-V||_{AC}. \tag{3.5}$$

On this basis when b/a is large the trial space of exponentials has to be capable of very close approximation to the solution if the method is to be used with confidence: although from the previous section we might expect this factor to be replaced by a mesh Péclet number, we shall not pursue these estimates further and instead we will consider below the accuracy attained at the nodes. One could also use these trial functions together with piecewise linear test functions: such a method is considered by Griffiths & Lorenz (1978), who show that this gives a lower error bound than any of the alternative upwind test functions (3.12) discussed in the next sub-section. When the coefficients a and b depend on x, local values of β can be used in each trial function and a similar error bound to (3.5) obtained.

(iii) Exponential test space (Hemker, 1977). This is motivated by some of the earliest work on superconvergence at the nodes, by de Boor & Swartz (1973) and Douglas & Dupont (1973). Consider a general one-dimensional problem, let $G^*_\xi(x)$ be the Green's function of the adjoint problem and denote the delta function $\delta(x-\xi)$ by $\delta_\xi(x)$. Then the weak form of the equation for G^*_ξ is

$$B(v,\ G^*_\xi) = (\delta_\xi, v) = v(\xi) \qquad \forall\ v \in H^1_0. \tag{3.6}$$

Now suppose U is a Petrov-Galerkin approximation to u obtained with a test space T^h so that

$$B(u-U,W) = 0 \qquad \forall\ W \in T^h. \tag{3.7}$$

Then (3.6) and (3.7) together give

$$u(\xi)-U(\xi) = B(u-U,G^*_\xi) = B(u-U,G^*_\xi-W) \qquad \forall\ W \in T^h, \tag{3.8}$$

and from (2.1a) we have

$$|u(\xi)-U(\xi)| \le C_1 ||u-U||_{AC} \inf_{W \in T^h} ||G^*_\xi - W||_{AC}. \tag{3.9}$$

As G^*_ξ has a discontinuous gradient at $x=\xi$, the last factor here will be reasonably small only when ξ is a mesh point. Then for any sensible choices of S^h and T^h the order of accuracy at the nodes should be double that in the $||\cdot||_{AC}$ norm: it should be noted, however, that for linear elements this is no improvement over the L_2 error bounds obtained by the Aubin-Nitsche arguments as in (2.43).

In the 1D model problem G^*_ξ consists of piecewise negative exponentials and hence, on a uniform mesh, we should take as test basis functions $\psi_j(x) = \psi(h^{-1}x-j)$, where

$$(1-e^{-\beta})\ \psi(t) = \begin{cases} 1-e^{-\beta(t+1)} & -1 \le t \le 0 \\ e^{-\beta t}-e^{-\beta} & 0 \le t \le 1 \ ; \end{cases} \tag{3.10}$$

these are the reflection of the trial functions given by (3.4) about $t=0$. Then G^*_ξ is approximated exactly and nodal values of u are exactly reproduced even for $-au'' + bu' = f(x)$, for general f, and any reasonable choice of trial space.

For a piecewise linear trial space, an alternative interpretation based on the error bounds of Section 2 is possible. From (2.17) and the subsequent argument, it is clear that $R^*T_0^h$ should be such as to approximate S_0^h well. An explicit expression for R was given in (2.49) for the operator in the model problem and in the $||\cdot||_{AC}$ norm. From this it is an easy calculation to show that indeed

$$\text{span}\ \{R^*\psi_j\} = \text{span}\ \{\phi_j\} \tag{3.11}$$

exactly, where the $\{\phi_j\}$ are the piecewise linear basis functions. Thus the resulting approximation is optimal in $||\cdot||_{AC}$ and hence exact at the nodes.

One of the disadvantages of using exponentials as either trial or test functions is the difficulty of evaluating the inner products involving these rapidly varying functions. Hemker (1977) has developed specialised quadrature formulae for this purpose. He also considered using these test functions only where the solution varied rapidly, as has Axelsson (1981) who used the very similar technique of introducing the negative exponential as a weight function in the bilinear form. More fundamental difficulties arise when any of these exponential-based techniques are extended into two dimensions and little progress has so far been reported.

3.2 Upwind methods

Zienkiewicz (1975) seems to have been the first to raise the possibility of choosing the test space in a Petrov-Galerkin scheme in order to obtain the same effects as upwind differencing. Mitchell and his colleagues quickly took up the challenge and a number of promising techniques were developed - see Christie et al. (1976), Heinrich et al. (1977) and the survey article Heinrich & Zienkiewicz (1979).

For the operator in the 1D model problem, it is apparent from the foregoing that either a positive exponential trial space in a Galerkin formulation or a negative exponential test space in a Petrov-Galerkin scheme will reproduce the Allen-Southwell difference operator: but clearly many other test spaces could achieve this. One of the simplest that may be used with a piecewise linear basis on a uniform mesh, $\{\phi_j\}$, is the following: with $\psi_j(x) = \psi(h^{-1}x-j)$ and $\sigma_j(x) = \sigma(h^{-1}x-j)$ we set

$$\psi(t) = \phi(t) + \alpha\sigma(t) \tag{3.12a}$$

with

$$\sigma(t) = \begin{cases} -3t(1-|t|) & |t| \le 1 \\ 0 & |t| > 1 \;. \end{cases} \tag{3.12b}$$

We see that $(\phi_j', \sigma_i') = 0$ for $1 \le i,\ j \le J-1$ so that the terms in the stiffness matrix arising from the diffusion operator do not depend on the parameter α: but for the convection terms we obtain

$$(\phi_j', \sigma_i) = -(\phi_j, \sigma_i') = \begin{cases} -\frac{1}{2} & j=i\pm1 \\ 1 & j=i \\ 0 & |j-1|>1 \;. \end{cases} \tag{3.13}$$

Thus for the 1D model problem we obtain

$$-a\delta^2 U_i + bh(\Delta_0 U_i - \tfrac{1}{2}\alpha\delta^2 U_i) = 0 \ . \tag{3.14}$$

Setting $\alpha=1$ gives the fully upwinded scheme of (1.18) and any choice such that $2a+\alpha bh>bh$, that is $\alpha>1-2a/bh$, avoids an oscillatory solution to the difference scheme by ensuring that it satisfies a maximum principle. The Allen & Southwell exponentially-fitted scheme is obtained by setting, as in (1.20)

$$\alpha = \xi := \coth(\tfrac{1}{2}bh/a) - (\tfrac{1}{2}bh/a)^{-1}$$
$$= \coth\tfrac{1}{2}\beta - 2/\beta \ . \tag{3.15}$$

It is easily seen that ξ varies smoothly from -1 to $+1$ as β ranges from $-\infty$ to $+\infty$, with $\xi \sim \beta/6$ as $\beta \to 0$ and $\xi \sim 1-2/\beta$ as $\beta \to \infty$.

It is interesting to see what choice of α is indicated by the error bounds in Theorem 2.2, and its Corollary 4, when the norm $||\cdot||_{AC}$ is used: a detailed analysis is given by Griffiths & Lorenz (1978). In Theorem 2.2 only C_2^M of (2.7) is affected by the choice of test functions, which should thus be chosen to maximise $||S^{-1}||$. Denoting by A and B the stiffness matrices representing the difference operators $-a\delta^2$ and $bh\Delta_0$, we see from the defining relation (2.11) that an expression for S^{-1} can be obtained from the following (we denote by $\underset{\sim}{V}$ the vector of nodal values of $V \in S_0^h$ and similarly for $W \in T_0^h$):

$$(1+3\alpha^2)A(\underset{\sim}{S}\underset{\sim}{V}) = [(1+\tfrac{1}{2}\alpha\beta)A + B]\underset{\sim}{V} \ . \tag{3.16}$$

The matrix on the left represents the inner product $(\cdot,\cdot)_{AC}$ in the basis (3.12a) of T_0^h, obtained using (3.13) and the fact that $(\sigma_j', \sigma_i') = 3(\phi_j', \phi_i')$. Then using Fourier analysis we find

$$C_2^M = \min \frac{||SV||_{AC}}{||V||_{AC}} = (1+3\alpha^2)^{-\frac{1}{2}}[(1+\tfrac{1}{2}\alpha\beta)^2 + \tfrac{1}{4}\beta^2\tan^2\tfrac{1}{2}\pi h]^{\frac{1}{2}}. \tag{3.17}$$

The maximum value is given quite accurately by neglecting the term $\tan^2\frac{1}{2}\pi h$, leading to the choice $\alpha = \beta/6$: this agrees with (3.15) for small β but at first sight seems quite unreasonable for large β.

Before considering this point further, let us derive the choice of α obtained by optimising the error bound in (2.17) : in particular, we choose α to minimise

$$\min_{\gamma} ||\gamma R^*\psi_j - \phi_j||^2_{AC}. \tag{3.18}$$

Using the expression for R in (2.49) and exploiting the fact that $\sigma' = -6(\phi-\frac{1}{2})$, it is a straightforward computation to obtain

$$||\gamma R^*\psi_j - \phi_j||^2_{AC} = (a/h)\int_{-1}^{1} \gamma^2[(1-\gamma^{-1})\phi' + (\beta-6\alpha)\phi + 3\alpha + \alpha\beta\sigma]^2 dt \tag{3.19}$$

and to find that this is minimised by

$$\alpha = \tfrac{5}{9}\beta(\beta^2+3)/(\beta^2+10). \tag{3.20}$$

This gives the correct behaviour, $\alpha \sim \beta/6$, for $\beta \to 0$ and very similar behaviour to that derived from (3.17), namely $\alpha \sim 5\beta/9$ for $\beta \to \infty$. Moreover, the latter is very easily understood in terms of the function fitting needed to minimise (3.19) : since ϕ and the constant 3α are the only even functions, one needs $\alpha = O(\beta)$; then $\gamma\alpha\beta\sigma$ is of very similar form to ϕ', the best fit being given by $\gamma\alpha\beta = 5/3$.

To understand the fact that these error bound arguments lead to $\alpha \to \infty$ as $\beta \to \infty$, instead of $\alpha \to 1$ as $\beta \to \infty$ in the Allen & Southwell scheme, we need to remember that they were based on the form of the problem (2.48) with homogenous boundary conditions and general data f. In the singular limit, the Allen & Southwell scheme drops the right-hand boundary condition and the data that goes with it and approximates $bu'=0$, $u(0)=0$ by $b\Delta_- U=0$, $U(0)=0$: and for a general piecewise linear data function F the Petrov-Galerkin method based on (3.12) with $\alpha=1$ will give

$$b\Delta_- U_i = \frac{h}{6}[6+\delta^2-3\Delta_0]F_i$$

$$\text{i.e.} \qquad b(U_i-U_{i-1}) = \frac{h}{12}[-F_{i+1}+8F_i+5F_{i-1}] \quad , \tag{3.21}$$

not a very convincing approximation to $bu'=F$. On the other hand with $\alpha \to \infty$, the scheme for $U^0=U-x$ with homogeneous boundary conditions at each end becomes

$$b\delta^2 U_i^0 = h\Delta_0(F-b). \tag{3.22a}$$

Constant data clearly gives a null solution and this takes the place of a boundary condition being dropped: for one integration can be effected and (3.22a) reduces for U to

$$b(U_i-U_{i-1}) = \tfrac{1}{2}h(F_i+F_{i-1}), \tag{3.22b}$$

a much more satisfactory approximation.

We should perhaps not consider these results for high β as too significant, for we have already seen that $||u_0-U_0^*||_{AC}$ for the optimal approximation U_0^* is very little reduced below $||u_0||_{AC}$. Thus, although (3.17) may seem heavily dependent on α, it is not surprising to find from (3.19) that

$$\min_{\gamma} ||\gamma R^*\psi_j-\phi_j||^2_{AC} = \left[1-\frac{(1+\frac{1}{2}\alpha\beta)^2}{1+3\alpha^2+\frac{1}{3}\beta^2+\frac{3}{10}\alpha^2\beta^2}\right]||\phi_j||^2_{AC} \tag{3.23}$$

which depends very little on α for large β. In the limit $\beta \to \infty$, the numerical factor in (3.23) quickly approaches 1/6 for any unbounded α and is 23/38 even for $\alpha=1$.

In addition to (3.12) with $\alpha=\xi$ and (3.10), any choice of test space that reproduces the exponentially-fitted Allen & Southwell scheme has the advantage that the corresponding discrete Green's function is exactly equal to that for the continuous problem with both arguments taken at node points: that is, in an obvious notation, $G_{jk}=G(jh,kh)$. Thus for the simplest such finite difference scheme applied to (2.48) for u_0, the nodal errors are given by

$$u_0(jh)-U_j^0 = \int_0^1 G(jh,y)f(y)dy - h\sum_{k=1}^{J-1} G(jh,kh)f(kh) \tag{3.24}$$

and are therefore wholly attributable to the trapezoidal rule applied to the integral of $G(jh,\cdot)f(\cdot)$. Similarly, for such a Petrov-Galerkin scheme the nodal errors are given by

$$u_0(jh)-U_j^0 = \sum_{k=1}^{J-1} E_{jk} \tag{3.25a}$$

where for instance for $k\geq j$,

$$E_{jk} = b^{-1}\frac{e^{\beta j}-1}{e^{b/a}-1}\int_0^1 [g_k(t)-g_k(0)\psi(t)-g_k(1)\psi(t-1)]f(kh+th)dt \tag{3.25b}$$

$$g_k(t) = e^{b/a-\beta(k+t)}-1. \tag{3.25c}$$

We can assume that $\psi(0)=1$, $\psi(\pm1)=0$ so that the kernel in (3.25b) is zero at the two ends of the range and the error depends on how well $\psi(t)$ matches $\exp(-\beta t)$ between these limits, (3.10) giving the perfect match.

In variable coefficient problems the choice of ψ, and in particular of the parameter α, can be made locally and similar error estimates to (3.25) derived. In two dimensions, precise error estimation and selection of ψ is considerably more difficult but the extension of (3.12) to bilinear elements on rectangles is straightforward: as in this case the trial basis functions are given by $\phi_{ij}(x,y) = \phi_i(x)\phi_j(y)$, the test functions can be taken as

$$\psi_{ij}(x,y)=[\phi_i(x)+\alpha_1\sigma_i(x)][\phi_j(y)+\alpha_2\sigma_j(y)] . \tag{3.26}$$

where (α_1,α_2) are chosen relative to the two components of $\underline{b}=(b_1,b_2)^T$ and the mesh spacing in the x and y directions. With quadrilaterals one can use such product functions of the isoparametric co-ordinates.

3.3 Streamline diffusion methods

As has been remarked previously, the Allen & Southwell scheme can be interpreted as having had extra diffusion added before the Galerkin method is used. Thus with a enhanced by $\frac{1}{2}\alpha bh$ piecewise linear elements reproduce (3.14) and $\alpha=\xi$ gives the Allen & Southwell scheme, but of course with ψ replaced by ϕ in any inhomogenous terms and in the error expressions (3.25). To extend this to two dimensions with a scalar diffusion would lead to excessive "cross-wind diffusion", that is normal to the direction of flow $\underline{b}$. Hughes & Brooks (1979, 1981) have therefore used in extensive computations a tensor diffusion given as follows: in (1.9) we replace $-\underline{\nabla}\cdot(a\underline{\nabla}u)$ by

$$-\underline{\nabla}\cdot(\underline{\underline{A}}\nabla u), \quad \text{where} \quad A_{\ell m}=a\delta_{\ell m}+\tilde{a}b_\ell b_m/|\underline{b}|^2. \tag{3.27}$$

On a uniform rectangular mesh with spacings h_1, h_2 the suggested choice of the parameter $\tilde{a}$ is

$$\tilde{a} = \tfrac{1}{2}(\xi_1 b_1 h_1 + \xi_2 b_2 h_2) \tag{3.28}$$

with $$\xi_m = \coth(\tfrac{1}{2} b_m h_m / a) - (\tfrac{1}{2} b_m h_m / a)^{-1}, \quad m = 1,2.$$

Though this choice is rather arbitrary it seems to work well in practice.

In their more recent paper, Hughes & Brooks have put this scheme into a Petrov-Galerkin framework by noting that

$$(\underline{\underline{A}}\nabla v, \underline{\nabla}\phi) = (a\underline{\nabla} v, \underline{\nabla}\phi) + (\underline{b}\cdot\underline{\nabla} v, (\tilde{a}/|\underline{b}|^2)\underline{b}\cdot\underline{\nabla}\phi) \quad . \tag{3.29}$$

Thus, assuming $\underline{\nabla}\cdot\underline{b} = 0$, the scheme is equivalent to using test functions

$$\psi_{ij} = \phi_{ij} + (\tilde{a}/|\underline{b}|^2)\underline{b}\cdot\underline{\nabla}\phi_{ij} \tag{3.30}$$

on just the convection term. For most trial spaces these functions will be discontinuous, which is quite acceptable for the convection term, but with $\psi_{ij} \notin H^1$ use of such test functions leads to consideration of so-called external approximations which is beyond the scope of these lectures. It is enough to note here, however, that if a is constant and U is bilinear then $\underline{\nabla}\cdot(a\underline{\nabla} U) = a\nabla^2 U = 0$ on each element. Hence, with the proviso that the term $(a\nabla^2 U, \underline{b}\cdot\underline{\nabla}\phi_{ij})$ is evaluated in this way the streamline diffusion method defined from (3.27) can be regarded as a Petrov-Galerkin method using test functions given by (3.30).

Alternatively, Johnson & Nävert (1981) have analysed a modification of this scheme in a way related to that followed in the next section. Starting from the reduced problem, (1.9) with $a = 0$, they use the fact that its solution also satisfies

$$(1-\delta\underline{b}\cdot\underline{\nabla})\ (\underline{b}\cdot\underline{\nabla} u + cu) = f - \delta\underline{b}\cdot\underline{\nabla} f. \tag{3.31}$$

Then the streamline diffusion method, with a modified right-hand side, is obtained by applying the Galerkin method to this equation with an appropriate choice of δ. They therefore obtain an error bound in a norm which depends on δ.

4. APPROXIMATE SYMMETRIZATION AND OPTIMAL APPROXIMATION

4.1 Motivation

The methods described in Section 3 were mainly motivated by the aim of high accuracy at nodal points or, equivalently for linear elements in one dimension, achieving a nearly optimal approximation in the $||\cdot||_{AC}$ norm. In addition, two of them involved a symmetrization of the problem: the Liouville transform did so directly; and, from the definition (2.3a) of the Riesz representer R and relation (3.11) which together imply

$$B(U,\psi_j) = (U, R^*\psi_j)_{AC} \propto (U,\phi_j)_{AC}, \tag{4.1a}$$

the use of exponential test functions leads to an approximation given by the symmetric system

$$(U,\phi_j)_{AC} = (f, R^{*-1}\phi_j) \qquad \forall\ \phi_j \in S_0^h . \tag{4.1b}$$

The adherence to the norm $||\cdot||_{AC}$ throughout the discussion of the constant coefficient model problem was deliberate, though most authors adopt the equivalent H_0^1 norm: it emphasises the derivation of the norm from the original problem and hints at its deficiencies when $c=0$ and the singular limit $a \to 0$ is approached.

These deficiencies were apparent in the very small reduction from $||u||_{AC}$ to $||u-U^*||_{AC}$ achieved by the optimal approximation U^*. An alternative interpretation is as follows: from the optimal approximation one wants to deduce as much information as possible about u, a problem in optimal recovery (see Micchelli & Rivlin, 1976); but for a sharp exponential boundary layer as in the model problem, the point value one mesh spacing inside the boundary gives very little information. The difficulty can also be attributed to the fact that the coefficient of the dominant convection term does not appear in the norm. Thus this term has its effect only in the rather awkward exponential which appears either in the test function or in the operator R^{*-1} appearing in (4.1b).

However, when $c=0$ in the problem (1.9), the operator can be factored and a symmetrization effected in an alternative way which has been exploited by Barrett & Morton (1980, 1981). Denoting the operators $\underline{\nabla}$ and $a\underline{\nabla}-\underline{b}$ by L_1 and L_2 respectively, the operator in (1.9a) is $L_1^*L_2$ and (4.1b) is based on the identity

$$(L_1R_1v, L_1w) = (L_1v, L_1R_1^*w) = (L_2v, L_1w) \qquad \forall\ v,w \in H_1, \tag{4.1c}$$

in which R_1 can be regarded as the Riesz representer of L_2 in a norm based on L_1 and defining the Hilbert space H_1. The alternative is to introduce a Riesz representer R_2^* for which

$$(L_2v, L_2R_2^*w) = (L_2v, L_1w) \qquad \forall\ v,w \in H_2 \tag{4.2a}$$

and which can be regarded as the Riesz representer of L_1 in a norm based on L_2 and defining a Hilbert space H_2: then an approximation U is generated from a test space $T_0^h = R_2^{*-1}S_0^h$ giving

$$(L_2U, L_2\phi_j) = (f, R_2^{*-1}\phi_j) \qquad \forall \phi_j \in S_0^h . \tag{4.2b}$$

Clearly if L_1w in (4.1c) spanned the same space as L_2v in (4.2a), we should have $R_2^* = R_1^{-1}$, $R_2^{*-1} = R_1$ so that in the model problem with R_1 given by (2.49) no exponentials would be involved. Unfortunately, such a relation does not hold exactly and some approximation is involved in aiming for optimality in the norm based on L_2 without the use of exponentials. In this section we consider how this is done first for problems in one dimension and then for those in two.

4.2 One dimensional problems

Consider the variable coefficient Dirichlet problem for u:

$$-(au')' + (bu)' = f \quad \text{on } (0,1) \tag{4.3a}$$

$$u(0) = g_L,\ u(1) = g_R . \tag{4.3b}$$

Then, from (2.25), we have for $v \in H^1$, $w \in H^1_0$

$$B(v,w) = (av',w') + ((bv)',w) = (av'-bv,w') . \tag{4.4}$$

We introduce a symmetric form with an arbitrary positive weighting function $\rho(x)$:

$$B_s(v,w) := (\rho a^2 v',w') + ([\rho b^2 + (\rho ab)']v,w) \tag{4.5a}$$

$$= (av' - bv, \rho[aw'-bw]) \qquad \forall\ v \in H^1,\ w \in H^1_0 . \tag{4.5b}$$

In addition to the usual assumptions on a and b, as in (2.23), we assume that ρ is normalised to have unit integral and is chosen so that on $(0,1)$ we have:

$$\text{(i)} \quad p(x) := \rho(x)a^2(x) > 0, \tag{4.6a}$$

$$\text{(ii)} \quad q(x) := \rho(x)b^2(x) + (\rho ab)'(x) \geq 0, \tag{4.6b}$$

$$\text{(iii)} \quad \alpha(x) := \rho(x)b(x) + (\rho a)'(x) \Rightarrow \alpha(x)b(x) \geq 0. \tag{4.6c}$$

This is easily achieved by, for instance, taking $(\rho a)' = 0$ where $b' \geq 0$ and $(\rho ab)' = 0$ where $b' < 0$; then $q \geq \rho b^2$ and $\alpha b \geq \rho b^2$. These assumptions ensure that $B_s(\cdot,\cdot)$ is a coercive form and that if $B(\cdot,\cdot)$ is coercive relative to the $||\cdot||_{AC}$ norm then it is also coercive relative to $||\cdot||^2_s := B_s(\cdot,\cdot)$.

Establishing the coercivity of $B(\cdot,\cdot)$ through (2.6) would require us to assume that there exists a $\delta > 0$ such that

$$(1-\delta)\int_0^1 av'^2 dx + \tfrac{1}{2}\int b' v^2 dx \geq 0 \qquad \forall\ v \in H^1_0. \tag{4.7}$$

However, we shall see in a moment that (2.1b,c) can be satisfied under much weaker conditions. Then we can apply either the Lax-Milgram lemma or Theorem 2.1 in respect of H_s, the Hilbert space formed from H^1_0 equipped with the $||\cdot||_s$ norm, and, retaining the notation of Barrett & Morton (1981), introduce a symmetrizing operator $N: H^1_0 \to H^1_0$ such that

$$B(v,Nw) = B_s(v,w) \qquad \forall\ v,w \in H^1_0 . \tag{4.8}$$

Indeed, it is not too difficult to construct N explicitly: we require from (4.4) and (4.5b) that

$$\int (av'-bv)\,[(Nw)' - \rho(aw' - bw)]dx = 0 \qquad \forall\ v \in H^1_0$$

and introduce $z = e^{-\lambda} v \in H^1_0$, where

$$\lambda(x) = \int_0^x (b/a)dt, \tag{4.9}$$

so that $av'-bv = az'e^{\lambda}$; then, as with (2.49), we have

$$(Nw)' = \rho(aw' - bw) + \text{const. } e^{-\lambda}/a \tag{4.10a}$$

from which a little manipulation gives

$$(Nw)(x) = (\rho aw)(x) + \int_x^1 (\alpha w - Ke^{-\lambda}/a)dy \tag{4.10b}$$

and the constant K is such that $(Nw)(0) = 0$. We have given the form which is appropriate for $\lambda > 0$, or $b \geq 0$, and it is also useful to note that N^+ its adjoint in the L_2 inner product is given by

$$(N^+f)(x) = (\rho af)(x) + \alpha(x)[F(x) - \tilde{F}] , \tag{4.11a}$$

where $F(x) := \int_0^x f(y)dy$

and $\tilde{F} \int_0^1 (e^{-\lambda}/a)dx := \int_{0_+}^1 (e^{-\lambda}/a)Fdx.$ (4.11b)

It is clear that N and N^+ involve an exponential kernel $e^{-\lambda}/a$ unless $\alpha \equiv 0$, which would require instead that ρ be proportional to the same exponential kernel. It is also clear from this construction why only the positivity of a is necessary to establish the hypotheses of Theorem (2.1): for if in $B(v,w)$ we set $w = e^{-\lambda}v$ we have

$$\left\{\sup_w \text{ or } \sup_v\right\} B(v,w) \geq B(e^{\lambda}w,w) = \int_0^1 ae^{\lambda}(w')^2dx. \tag{4.12}$$

An optimal approximation to u in the norm $||\cdot||_s$ can now be constructed using the symmetrizing operator N. If the trial space S^h is spanned by $\{\phi_j\}$, taking the test space as $T_0^h = NS_0^h$ in a Petrov-Galerkin method gives $U^* \in S_E^h$ such that

$$B(U^*,N\phi_j) = (f,N\phi_j) \qquad \forall\ \phi_j \in S_0^h. \tag{4.13}$$

Subtracting from a similar equation for u and using (4.8) establishes the optimality of U^*,

$$B_s(u - U^*,\phi_j) = 0 \qquad \forall\ \phi_j \in S_0^h . \tag{4.14}$$

It is important to note too that the discrete equations for U^* only involve the operation of N and N^+ on the data and the test functions never need to be obtained explicitly: if ϕ_0 and ϕ_J are the basis functions corresponding to the data g_L and g_R on the left and right respectively, we have from (4.12) and (4.8)

$$B_s(U^*-g_L\phi_0-g_R\phi_J,\phi_j) = (N^+f,\phi_j) - B(g_L\phi_0+g_R\phi_J,N\phi_j) \qquad \forall\ \phi_j \in S_0^h . \tag{4.15a}$$

This in turn can be reduced by (4.10) to

$$B_s(U^*,\phi_j) = (N^+f,\phi_j) - \alpha_j \ell_\lambda(g_L\phi_0+g_R\phi_J) \qquad \forall\ \phi_j \in S_0^h . \tag{4.15b}$$

where
$$\alpha_j := \int_0^1 \alpha\phi_j dx \tag{4.16}$$

and
$$\ell_\lambda(w) \int_0^1 (e^{-\lambda}/a)dx := \int_0^1 (e^{-\lambda}/a)(aw'-bw)dx. \tag{4.17}$$

In this form we can see that the exponential kernel is involved only in the calculation of the averages $\tilde{F}$, $\ell_\lambda(\phi_0)$ and $\ell_\lambda(\phi_J)$. We can also regard the equation for U^* as obtained by operating on (4.3) with the symmetrizing operator N^+ and then using the Galerkin method: this can then be compared with the streamline diffusion method in the form (3.31).

In their consideration of one dimensional problems, Barrett & Morton (1980, 1981) eschew exponentials completely by approximating the averages (4.11b) and (4.17) by a weighting function $\varepsilon(x)$, normalised to unit integral, or a delta-function at $x = 0$: they denote the corresponding operator (4.10) by N_ε or N_0 and the corresponding linear functional (4.17) by ℓ_ε or ℓ_0. In the former case $N_\varepsilon : H_0^1 \to H_0^1$ gives a proper Petrov-Galerkin method with $T_0^h = N_\varepsilon S_0^h$: but the symmetrization (4.8) is not exactly achieved and instead of (4.14) we have for the approximation U^N,

$$B_s(U^N,\phi_j) + \alpha_j \ell_\varepsilon(U^N) = (N_\varepsilon^+ f,\phi_j) \qquad \forall\ \phi_j \in S_0^h . \tag{4.18}$$

When the delta function is used N_0 is defined by (4.10) with $K = 0$, for no value will ensure that $(N_0 w)(0) = 0$ if $w(0) = 0$, and N_0^+ by (4.11a) with $\tilde{F} = 0$. Thus the resulting method is not strictly of Petrov-Galerkin form but the approximation still satisfies equation (4.18), with ℓ_ε and N_ε^+ replaced by ℓ_0 and N_0^+, and indeed is the simplest to use and the most appropriate in the singular limit $b/a \to \infty$.

Introducing $V^* \in S_0^h$ such that

$$B_s(V^*,\phi_j) = \alpha_j \qquad \forall\ \phi_j \in S_0^h , \tag{4.19}$$

Barrett & Morton (1981) show that for a problem with no turning points, $b(x) > 0$, U^N is uniquely determined if ε is chosen to ensure that $1+\ell_\varepsilon(V^*) \neq 0$ and

$$\begin{aligned} ||u-U^N||_s^2 &= ||u-U^*||_s^2 + ||U^*-U^N||_s^2 \\ &= ||u-U^*||_s^2 + \left[\frac{||V^*||_s}{1+\ell_\varepsilon(V^*)}\right]^2 [\ell_\varepsilon(u-U^*)]^2 . \end{aligned} \tag{4.20}$$

This estimate also holds for problems with a homogeneous Neumann condition at $x = 1$ but for the Dirichlet problem it is easy to show that $||V^*||_s \le 1$. The same result holds using N_0 and ℓ_0 and in that case the authors show that under quite general conditions $\ell_0(V^*) \ge 0$. Precise error bounds may then be derived: thus we have

$$||u-U^N||_s^2 \le ||u-U^*||_s^2 + a^2(0)[u'(0) - U^{*\prime}(0)]^2 \tag{4.21}$$

and, for example with constant coefficients and linear elements on a uniform mesh,

$$|U_j^N - U_j^*| \le \frac{3}{2}\frac{a}{b}\left|\frac{U_1^N-U_0^N}{h} - \frac{f(0)}{b}\right| + \frac{3}{2b}\left|\int_0^1 e^{-bx/a}\,[f(x)-f(0)]dx\right| + O(e^{-b/a}). \tag{4.22}$$

Similar results are given for variable coefficient problems. They show that when b/a or $\lambda(1)$ is large and $f(x)$ is nearly constant near $x = 0$ then U^N is very close to U^*, the optimal approximation in $||\cdot||_s$. Even some turning-point problems can be approximated well in this way. For a single turning point at ξ with $b(x) \le 0$ for $x \le \xi$ and $b(x) \ge 0$ for $x \ge \xi$, the delta function is placed at ξ and comparable error bounds obtained.

To compare these methods with other Petrov-Galerkin methods, consider the constant coefficient problem with ρ also taken constant: then from (4.10) the test functions are given by

$$\psi_j(x) = (N_\varepsilon \phi_j)(x) = a\phi_j(x) + b\int_x^1 (\phi_j - \overline{\phi}_j \varepsilon)dy. \tag{4.23}$$

These are not localised functions though linear combinations of successive pairs can be localised. The choice $\varepsilon(x) \equiv 1$ corresponds to the H^{-1} least squares formulation of Bristeau et al. (1980) which, as Barrett (1980) shows, is not very accurate for the simple model problem when bh/a is moderately large. On the other hand, the localised upwind test functions (3.12) can also be related to (4.23): we have already noted that $\sigma' = 6(\overline{\phi} - \phi)$, so that these test functions can be written as

$$\psi_j(x) = \phi_j(x) + (6\alpha/h)\int_x^1 (\phi_j - \overline{\phi}_j/2h)dy. \tag{4.24}$$

That is, they correspond to taking a different weighting function for each ϕ_j, equal to $(1/2h)$ over the support of ϕ_j, if the parameter α is taken as $\alpha = \beta/6$

To conclude this section we consider a numerical example together with the resulting recovery problem. The example, taken from Barrett (1980), is for

$$-10^{-3}u'' + [(1.0-0.98x)u]' = 0, \qquad u(0) = 1 \tag{4.25}$$

with either the Neumann condition $u'(1) = 0$ or the Dirichlet condition $u(1) = 49.95$ which gives the same solution. The table below gives various approximations for each case using piecewise linear elements on a mesh with $h = 0.1$; only the last three nodal values are given. The Galerkin approximation is U^G and U^U is the Petrov-

	Dirichlet case			Neumann case		
Node j	8	9	10	8	9	10
$u(jh)$	4.73	9.31	49.95	4.73	9.31	49.95
U_j^G	5.73	18.41	"	3.85	1.76	25.34
U_j^U	4.70	5.71	"	4.70	5.70	50.03
U_j^E	3.57	5.35	"	3.57	5.35	12.64
U_j^N	4.75	7.01	"	4.85	6.77	45.29
$u_R(jh)$	4.57	9.83	"	4.85	8.60	50.03

TABLE Results from approximating (4.25) with linear elements.

Galerkin result using the upwind scheme of Heinrich et al. (1977), that is with test functions (3.12), and α given the nodal values of $\alpha_{crit} = 1-2/\beta$. The row labelled U^E corresponds to test functions $\psi_j = \phi_j e^{-\lambda}$, the case of exact symmetrization obtained from (4.10b) with $\alpha = 0$, $\rho \propto e^{-\lambda}$: this also corresponds to using an exponential weighting in the inner product as in Axelsson (1981). The penultimate row gives U^N obtained from (4.18) using N_0 and $\rho \equiv 1$ in the Dirichlet case but $\rho(x) = (1.0-0.98x)^{-1}$ in the Neumann case in order to give more weight to the right of the interval.

None of these nodal values is particularly accurate, except $U^U(1)$ in the Neumann case which results from U^U satisfying a simple flux conservation relation. However U^N does not purport to have accurate nodal values: it aims instead at being a nearly optimal fit to u in the $||\cdot||_s$ norm, which is close to a weighted L_2 norm in this case. This optimality property can be combined with any further a priori knowledge of u, such as smoothness, monotonicity, positivity, etc., to give more accurate estimates for u. For the present problem, in the Dirichlet case, we expect u to be well approximated by an exponential in the boundary layer near $x = 1$, of the form

$$u_R(x) = \lambda_1 e^{\lambda_2(x-1)} + \lambda_3 \qquad (4.26)$$

for some constants λ_1, λ_2 and λ_3. One equation for determining these parameters is provided by the boundary condition $u_R(1) = 49.95$ and the other two can be obtained by assuming that locally U^N is a best fit to u_R and hence

$$B_s(u_R - U^N, \phi_j) = 0 \qquad j = J-2,\ J-1. \qquad (4.27)$$

The result of this procedure is given in the table: it can be seen that the value for $u(0.9)$ is accurate to 5%; and the boundary layer half-width can be predicted as 0.0394 as compared with the exact value 0.0447.

In the Neumann case, in order to satisfy the boundary condition we take

$$u_R(x) = \lambda_1 e^{\lambda_2(x-1)^2} + \lambda_3. \qquad (4.28)$$

One equation for the parameters is obtained from integrating (4.25) over $(x,1)$ to obtain a relation of the form

$$-au' + bu = \text{const.} = b(1)u(1), \qquad (4.29)$$

and others can be obtained from equations of the form (4.27), including $j = J$. However, if only $u(1)$ is required, substitution from (4.29) directly into $B_s(u-U^N, \phi_J) = 0$ provides a good approximation: this is the value given in the table, together with interior values obtained for (4.28) and (4.27).

4.3 Two-dimensional problems

Before extending these techniques of approximate symmetrization to two dimensions, it is useful to place them more carefully in the abstract framework of Section 2. Leaving aside for the moment the case when $\varepsilon(x)$ is a delta function, we can work in H_s, that is H_0^1 equipped with the $||\cdot||_s$ norm, and define N_ε so that $T_0^h = N_\varepsilon S_0^h \subset H_s$. For the approximation U^N given by

$$B(U^N,W) = (f,W) = B(u,W) \qquad \forall\ W\in T_0^h\ , \tag{4.30}$$

we have, from defining R^* as in (2.3a),

$$B_s(u-U^N,R^*W) \equiv B(u-U^N,W) = 0 \qquad \forall\ W\in T_0^h\ . \tag{4.31}$$

Suppose now that the constant $\Delta\in[0,1)$ is such that

$$\inf_{W\in T_0^h} ||V-R^*W||_s \le \Delta\ ||V||_s \qquad \forall\ V\in S_0^h\ . \tag{4.32}$$

Then with U^* given by (4.14) and repeating the argument following (2.17), we have

$$\begin{aligned} ||U^*-U^N||_s^2 &= B_s(u-U^N,U^*-U^N) \\ &= B_s(u-U^N,U^*-U^N-R^*W) \qquad \forall\ W\in T_0^h \\ &\le \Delta||u-U^N||_s||U^*-U^N||_s\ . \end{aligned} \tag{4.33}$$

Thus from $||u-U^N||_s^2 = ||u-U^*||_s^2 + ||U^*-U^N||_s^2$ we obtain

$$||u-U^N||_s \le (1-\Delta^2)^{-\frac{1}{2}}||u-U^*||_s\ . \tag{4.34}$$

It is clear that a good approximation is obtained if in particular R^*N_ε is close to the identity: that is, comparing (4.31) to (4.8), we take N_ε to approximate $R^{*-1} = N$.

When using N_0 by taking $\varepsilon(x)$ as a delta function, we generate test functions W which for non-turning point problems are not in H_0^1 and for turning point problems may not even be in H^1: in two dimensions the corresponding test functions would not be zero on the inflow Dirichlet boundary $\Gamma_D\cap\Gamma^-$. Though it may be possible to extend the definition of R^* to such functions and hence to establish an approximation result like (4.32), it is difficult to do this so as to maintain (4.31). For $B(v,w)$ to be defined by (2.25) so that (2.24) holds one needs to have $\nabla^2v\in L_2(\Omega)$: thus as a minimum in (4.31) one needs to assume greater smoothness on u; and to define R^* by (4.31) using the Riesz representation theorem one needs to work in smoother spaces than H_s. We shall therefore regard the approximation derived using N_0 as a limiting case of those obtained from N_ε. In the one dimensional problems treated above this mainly required establishing a uniform bound on $||K^{-1}||$ as $\ell_\varepsilon \to \ell_0$, where K is the stiffness matrix for the system in (4.18).

We consider then two dimensional problems covered by Theorem 2.3, with the added restriction that $c \equiv 0$, and in the space H_s with the definition of $B_s(\cdot,\cdot)$ extended from (4.5) to

$$B_s(v,w) := (\rho a^2 \underline{\nabla} v, \nabla w) + ([\rho b^2 + \underline{\nabla}\cdot(\rho a\underline{b})]v,w) \tag{4.35a}$$

$$= (a\underline{\nabla} v - \underline{b}v, \rho(a\underline{\nabla} w - \underline{b}w)) + (\underline{n}\cdot\underline{b}v, \rho aw)_{\Gamma_N} \quad \forall\ v\epsilon H^1, w\epsilon H^1_0. \tag{4.35b}$$

The assumptions made in (4.6) are generalised in a natural way with $\underline{\alpha} := \rho\underline{b} + \underline{\nabla}(\rho a)$ and $\underline{\alpha}\cdot\underline{b} \geq 0$. Then if all these coefficients are sufficiently smooth $B(\cdot,\cdot)$ satisfies the hypotheses of Theorem 2.1 in $H_s \times H_s$. Thus there exists a symmetrizing operator N satisfying (4.8) which requires explicitly that

$$(a\underline{\nabla} v-\underline{b}v, [\underline{\nabla}(Nw) - \rho(a\underline{\nabla} w-\underline{b}w)]) + (\underline{n}\cdot\underline{b}v, [Nw-\rho aw])_{\Gamma_N} = 0 \qquad \forall\ v\epsilon H^1_0 . \tag{4.36}$$

In case that the vector field $\underline{b}/a$ is irrotational a scalar λ can be introduced, as in (4.9), such that $\nabla\lambda = \underline{b}/a$ and with $\lambda = 0$ at some inflow point. Then introducing $z = e^{-\lambda}v$ and using the divergence theorem, the problem for Nw becomes

$$\underline{\nabla}\cdot(ae^{\lambda}\underline{\nabla}(Nw)) = \underline{\nabla}\cdot[a\rho e^{\lambda}(a\underline{\nabla} w-\underline{b}w)] \quad \text{in } \Omega \tag{4.37a}$$

$$a\frac{\partial}{\partial n}(Nw) + \underline{n}\cdot\underline{b}(Nw) = \rho a^2\frac{\partial w}{\partial n} \quad \text{on } \Gamma_N , \quad Nw = 0 \quad \text{on } \Gamma_D . \tag{4.37b}$$

This is not particularly useful as a starting point for approximating $N\phi_j$ to obtain test functions in a Petrov-Galerkin method. One could introduce a stream function $\psi(x,y)$ to take the place of the constant in (4.10) and for which one would then have

$$\partial_x(Nw) = \rho(a\partial_x w - b_1 u) + (e^{-\lambda}/a)\partial_y\psi \tag{4.38a}$$

$$\partial_y(Nw) = \rho(a\partial_y w - b_2 u) - (e^{-\lambda}/a)\partial_x\psi , \tag{4.38b}$$

with the boundary conditions on ψ obtained from (4.37b). However, such an approach has not so far been followed up directly, though it does motivate one of the two approaches that have been used.

This, the most direct extension of the one-dimensional technique, was reported in Morton & Barrett (1980). As with the upwind method based on (3.26), it uses bilinear elements on rectangles and correspondingly generates test functions given by

$$(N_\varepsilon\phi_{ij})(x,y) = (N_\varepsilon^{(x)}\phi_i)(x).(N_\varepsilon^{(y)}\phi_j)(y), \tag{4.39}$$

where $N_\varepsilon^{(x)}$ is as in (4.23) but based on $a(x,y_j)$ and $b_1(x,y_j)$ with $N_\varepsilon^{(y)}$ defined similarly. The results for standard test problems in which $\underline{b}$ has a fixed direction are quite good but the method is not adequate when $\underline{b}$ corresponds to very curved flow lines.

An alternative approach has been given in Barrett & Morton (1981). Corresponding to (4.14), (4.15) the exact solution u is easily seen to satisfy

$$B_s(u,\phi_j) = (\rho a f,\phi_j) + (\underline{b}u - a\nabla u, \underline{\alpha}\phi_j) \qquad \forall\ \phi_j \in S_0^h . \qquad (4.40)$$

Thus suppose we introduce the flux function $\underline{v} = \underline{b}u - a\nabla u$ and approximate it by $\underline{V}$. Then an approximation $U \in S_E^h$ to u can be obtained from

$$B_s(U,\phi_j) = (\rho a f,\phi_j) + (\underline{V}, \underline{\alpha}\phi_j) \qquad \forall\ \phi_j \in S_0^h \qquad (4.41)$$

and any approximation scheme for $\underline{V}$. Defining U^* now by (4.14) one obtains

$$B_s(U^*-U,\phi_j) = (\underline{\alpha}\cdot(\underline{v}-\underline{V}),\phi_j) \qquad \forall\ \phi_j \in S_0^h . \qquad (4.42)$$

Moreover, if we define s as $\underline{\alpha}\cdot\underline{v}$ and denote by S^* its best L_2 fit in S^h such that $s = S^*$ on Γ_D, then

$$B_s(U^*-U,\phi_j) = (S^* - \underline{\alpha}\cdot\underline{V},\phi_j) \qquad \forall\ \phi_j \in S_0^h$$

and hence

$$||U^*-U||_s^2 \le |||\underline{\alpha}|^{-1}S^* - \hat{\underline{\alpha}}\cdot\underline{V}||\cdot|||\underline{\alpha}|(U^*-U)||,$$

where $\hat{\underline{\alpha}} = \underline{\alpha}/|\underline{\alpha}|$. Introducing the constant γ such that

$$|\rho\underline{b} + \underline{\nabla}(\rho a)|^2 \le \gamma[\rho b^2 + \underline{\nabla}\cdot(\rho\underline{b}a)] \qquad (4.43)$$

we obtain the following relationship between the deviation of U and V from their "optimal" approximations:

$$||U^*-U||_s \le (\gamma/|\underline{\alpha}|)||S^* - \underline{\alpha}\cdot\underline{V}|| . \qquad (4.44)$$

Thus $\underline{V}$ should be constructed by approximating the equation $\underline{\nabla}\cdot\underline{v} = f$ in such a way that $\underline{\alpha}\cdot\underline{V}$ is close to S^*: boundary conditions can be obtained by setting $\underline{V} = \underline{b}U - a\underline{\nabla}U$ on the inflow boundary and U and $\underline{V}$ obtained from an alternating iterative procedure. One such scheme was given in Barrett & Morton (1981) but what is the most effective scheme is not yet clear.

To conclude, we believe that some sort of symmetrization is the most useful basic approach to approximating non-self-adjoint problems by finite element methods. Other more economic methods of adequate accuracy may then be derived from these. In Section 3 use of the exponential test functions turned out to be such a basic approach, based on the symmetric part of the operator L being used to define the norm $||\cdot||_{AC}$: then the Allen & Southwell difference operator can be regarded as a practical shortcut to forming the stiffness matrix and various other upwinded test functions as approximating the Green's function in (3.25) in order to model the effect of the inhomogeneous data f. In this last section a natural alternative symmetrization based on the norm $||\cdot||_s$ has been presented. It gives quite a different type of approximation, much less closely linked to finite difference methods, and can yield so-called sub-gridscale information. The further development of these two approaches for two-dimensional problems should show which is the more useful.

It is a pleasure to acknowledge the valuable discussions with Dr. J.W. Barrett that have taken place during and prior to the preparation of these lecture notes.

5. REFERENCES

[1] Agmon, S., Douglis, A., and Nirenberg, L, 1964. Estimates near the boundary for solutions of elliptic partial differential equations satisfying general boundary conditions II. Comm. Pure Appl. Math, 17, 35-92.

[2] Allen, D., and Southwell, R., 1955. Relaxation methods applied to determine the the motion, in two dimensions, of a viscous fluid past a fixed cylinder. Quart. J. Mech. and Appl. Math., VIII, 129-145.

[3] Axelsson, O., 1981. Stability and error estimates of Galerkin finite element approximations for convection-diffusion equations. I.M.A. J. Numer. Anal. 1, 329-345.

[4] Babuška, I., and Aziz, A.K., 1972. Survey lectures on the mathematical foundations of the finite element method. The Mathematical Foundations of the Finite Element Method with Applications to Partial Differential Equations (ed. A.K. Aziz), New York: Academic Press, 3-363.

[5] Barrett, J.W., 1980. Optimal Petrov-Galerkin methods. Ph.D. Thesis, University of Reading.

[6] Barrett, J.W. and Morton, K.W., 1980. Optimal finite element solutions to diffusion-convection problems in one dimension. Int. J. Num. Meth. Engng. 15, 1457-1474.

[7] Barrett, J.W., and Morton, K.W., 1981. Optimal Petrov-Galerkin methods through approximate symmetrization. I.M.A. J. Numer. Anal. 1, 439-468.

[8] Barrett, J.W., and Morton, K.W., 1981. Optimal finite element approximation for diffusion-convection problems. Proc. MAFELAP 1981 Conf. (ed. J.R. Whiteman).

[9] Barrett, K.E., 1974. The numerical solution of singular-perturbation boundary-value problems. J. Mech. Appl. Math., 27, 57-68.

[10] Barrett, K.E., 1977. Finite element analysis for flow between rotating discs using exponentially weighted basis functions. Int. J. Num. Meth. Engng., 11, 1809-1817.

[11] de Boor, C., and Swartz, B., 1973. Collocation at Gaussian points. SIAM J. Numer. Anal., 10, 582-606.

[12] Bristeau, M.O., Pirronneau, O., Glowinski, R., Periaux, J., Perrier, P., and Poirier, G., 1980. Application of optimal control and finite element methods to the calculation of transonic flows and incompressible viscous flows. I.M.A. Conf. Numerical Methods in Applied Fluid Dynamics (ed. B. Hunt), Academic Press, 203-312.

[13] Christie, I., Griffiths, D.F., Mitchell, A.R., and Zienkiewicz, O.C., 1976. Finite element methods for second order differential equations with significant first derivatives. Int. J. Num. Meth. Engng., 10, 1389-1396.

[14] Ciarlet, P.G., 1978. The Finite Element Method for Elliptic Problems. North-Holland (Amsterdam).

[15] Doolan, E.P., Miller, J.J.H., and Schilders, W.H.A., 1980. Uniform Numerical Methods for Problems with Initial and Boundary Layers. Dublin: Boole Press.

[16] Douglas, J., Jr., and Dupont, T., 1973. Superconvergence for Galerkin methods for the two point boundary problem via local projections. Numer. Math., 21, 270-278.

[17] Gresho, P.M., and Lee, R.L., 1979. Don't suppress the wiggles - they're telling you something. Finite Element Methods for Convection Dominated Flows (ed. T.J.R. Hughes) AMD Vol. 34, Am. Soc. Mech. Eng., 37-61.

[18] Griffiths, D., and Lorenz, J, 1978. An analysis of the Petrov-Galerkin finite element method. Comp. Meth. Appl. Mech. Engng., 14, 39-64.

[19] Guymon, G.L., Scott, V.H., and Herrmann, L.R., 1970. A general numerical solution of the two-dimensional diffusion-convection equation by the finite element method. Water Resources 6, 1611-1617.

[20] Guymon, G.L., 1970. A finite element solution of a one-dimensional diffusion-convection equation. Water Resources 6, 204-210.

[21] Heinrich, J.C., Huyakorn, P.S., Mitchell, A.R., and Zienkiewicz, O.C., 1977. An upwind finite element scheme for two-dimensional convective transport equations. Int. J. Num. Meth. Engng., 11, 131-143.

[22] Heinrich , J.C., and Zienkiewicz, O.C., 1979. The finite element method and 'upwinding' techniques in the numerical solution of convection dominated flow problems. Finite Element Methods for Convection Dominated Flows (ed. T.J.R. Hughes) AMD Vol. 34, Am. Soc. Mech. Eng., 105-136.

[23] Hemker, P.W., 1977. A numerical study of stiff two-point boundary problems. Thesis, Amsterdam: Math. Cent.

[24] Hughes, T.J.R., and Brooks, A., 1979. A multidimensional upwind scheme with no crosswind diffusion. Finite Element Methods for Convection Dominated Flows (ed. T.J.R. Hughes) AMD Vol. 34, Am. Soc. Mech. Eng., 19-35.

[25] Hughes, T.J.R., and Brooks, A., 1981. A theoretical framework for Petrov-Galerkin methods with discontinuous weighting functions: application to the streamline-upwind procedure. To appear in Finite Elements in Fluids Vol. 4 (ed. R.H. Gallagher), J. Wiley & Sons : New York.

[26] Johnson, C., and Nävert, U., 1981. An analysis of some finite element methods for advection-diffusion problems. Conf. on Analytical and Numerical Approaches to Asymptotic Problems in Analysis (eds. O. Axelsson, L.S. Frank and A. van der Siuis), North-Holland.

[27] Lesaint, P., and Zlamal, M., 1979. Superconvergence of the gradient of finite element solutions. R.A.I.R.O. Numer. Anal., 13, 139-166.

[28] Micchelli, C.A., and Rivlin, T.J., 1976. A survey of optimal recovery. Optimal Estimation in Approximation Theory (ed. C.A. Micchelli & T.J. Rivlin), Plenam Press : New York.

[29] Morton, K.W., and Barrett, J.W., 1980. Optimal finite element methods for diffusion-convection problems. Proc. Conf. Boundary and Interior Layers - Computational and Asymptotic Methods (ed. J.J.H. Miller), Boole Press : Dublin, 134-148.

[30] Oden, J.T., and Reddy, J.N., 1976. An Introduction to the Mathematical Theory of Finite Elements, Wiley-Interscience : New York.

[31] Strang, G., and Fix, G.J., 1973. An Analysis of the Finite Element Method, Prentice Hall : New York.

[32] Thomée, V, and Westergren, B., 1968. Elliptic difference equations and interior regularity. Numer. Math. II, 196-210.

[33] Zienkiewicz, O.C., Gallagher, R.H. and Hood, P., 1975. Newtonian and non-Newtonian viscous incompressible flow, temperature induced flows : finite element solution. 2nd Conf. Mathematics of Finite Elements and Applications (ed. J.R. Whiteman), London : Academic Press.

[34] Zienkiewicz, O.C., 1977. The Finite Element Method, London : McGraw Hill.

[35] Zlamal, M., 1977. Some superconvergence results in the finite element method. Mathematical Aspects of Finite Element Methods. Springer-Verlag.

[36] Zlamal, M., 1978. Superconvergence and reduced integration in the finite element method. Math. Comp. 32, 663-685.

AN INTRODUCTION TO PIECEWISE-LINEAR HOMOTOPY ALGORITHMS FOR SOLVING SYSTEMS OF EQUATIONS*

Michael J. Todd

Abstract. We describe a class of algorithms known as piecewise-linear homotopy methods for solving certain (generalized) zero-finding problems. The global and local convergence properties of these algorithms are discussed. We also outline recent techniques that have been proposed to improve the efficiency of the methods.

1. Introduction

This paper is intended to introduce the reader to a class of algorithms for solving certain (generalized) zero-finding problems. Thus, given a continuous function $f: R^n \to R^n$, we may wish to find $x^* \in R^n$ with

$$0 = f(x^*). \tag{1.1}$$

Since this is such a general and all-embracing problem, it is worthwhile describing some particular instances of (1.1) that motivated the early developers of these methods.

The first algorithm of the class we shall consider was devised by Scarf [69] in 1967. Scarf's method gave a constructive proof of Brouwer's fixed-point theorem, that any continuous function g mapping an n-dimensional simplex S into itself has a fixed point. While this is a very general result, Scarf was motivated by the desire to compute an equilibrium price vector for a model of an economy. His algorithm was derived from an earlier method he had developed to prove constructively the non-emptiness of the core in certain n-person games--see [68]. The novel and combinatorial proof of convergence was inspired by a similar argument of Lemke and Howson [53] (in a paper that originated the closely related field of complementary pivot theory--see Lemke [52] for a survey). From 1967-1972, those who worked in the field all came from mathematical economics and game theory (Scarf, Hansen, Kuhn and Shapley) or operations research and optimization theory (Eaves, Merrill and Saigal). Their motivation was the computation of fixed points that arise in various economic and game-theoretic models and the computation of

*Research partially supported by the National Science Foundation under grant ECS-7921279 and by a fellowship from the Alfred P. Sloan Foundation.

stationary points in optimization. For instance, if g maps R^n continuously into an n-simplex S, we may set $f(x) = x-g(x)$ so that solutions to (1.1) are fixed points of g. If we seek the unconstrained minimizer of a continuously differentiable function $\theta: R^n \to R$, we may set f equal to the gradient of θ; then solutions to (1.1) are stationary points of θ.

Consideration of very natural generalizations in these models--e.g., the introduction of production into economic equilibrium problems or constraints into optimization problems--led to the need to compute solutions to

$$0 \in F(x^*), \tag{1.2}$$

where F is an upper semi-continuous mapping from R^n into nonempty compact convex subsets of R^n. (This notion of continuity requires that $\{x \in R^n: F(x) \subseteq V\}$ be open for each open $V \subseteq R^n$.) There is a corresponding fixed-point theorem for such mappings, due to von Neumann and Kakutani, that guarantees the existence of a solution to (1.2) under suitable conditions. Remarkably, the algorithms could easily be adapted to this more general problem.

These general problems and their origins forced certain features on the algorithms devised to solve them. Since it was desired to handle instances of (1.2) as well as of (1.1), no advantage could be taken of smoothness in the basic algorithm. In addition, because the forms of function f that could arise from economic models was very general, the algorithms had to guarantee convergence for any f (or F) satisfying suitable boundary conditions. Initially, these boundary conditions were basically that $f(x) = x-g(x)$ for some g mapping a simplex S into itself. Thus the algorithms were called fixed-point methods. Another term that has been used is simplicial or simplicial approximation methods. It now seems more natural to state the problem in terms of f, and because of the mode of operation of recent algorithms we shall call them piecewise-linear homotopy methods. They are closely related to early embedding methods and more recent continuation methods, as we shall see in section 6. The newer algorithms retain the properties of global convergence under very weak boundary conditions that hold naturally in several applications and of being able to handle point-to-set mappings. In addition, techniques have been devised to take advantage of smoothness in problems of form (1.1). Nevertheless, where a classical method is available for such a problem and a good approximation to a zero is known, such a method is likely to be more efficient than a piecewise-linear homotopy algorithm.

It is time to illustrate the basic ideas of this class of algorithms. Suppose we wish to find a fixed point of a continuous function $g: [0,1]^n \to [0,1]^n$ and suppose further that g maps the cube $[0,1]^n$ into its interior. Let us define $f(x) = x-g(x)$. Then $f_i(x)$ is negative (positive) when x_i is 0 (1): this is our boundary condition in this case. Consider first a typical example of an early algorithm. Such a method finds an exact zero of a function $\tilde{f}$ approximating f, for example, a piecewise-linear (PL) approximation: see figure 1.1 when $n = 1$.

In order to obtain a sequence converging to a zero of f, a suitable sequence $f^1, f^2, \ldots,$ of functions is chosen, e.g., PL approximations with respect to finer and finer meshes. The drawback of these early methods is that all information gained in finding a zero of f^k is wasted in searching for a zero of f^{k+1}. Thus the methods are typically applied once with an appropriately fine mesh. These techniques are therefore somewhat inefficient.

For a 1-dimensional problem as in figure 1.1, the methods proceed from left to right until an interval containing a zero of f^k is found. We can also

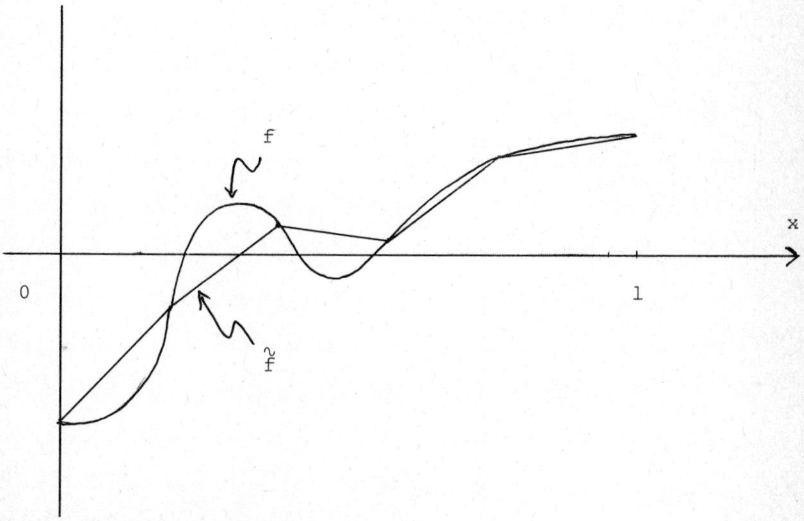

Figure 1.1

illustrate the later algorithms that introduce an extra dimension for such a problem. These can be classified as restart (Merrill [55], Kuhn and McKinnon [44]) or continuous deformation (Eaves [13], Eaves and Saigal [15]) methods.

The restart methods deform a simple (usually linear) function r into the function f of interest using an extra homotopy parameter t and trace the zeroes of this deformation. More precisely, $r^1 \equiv r$ is deformed into a (coarse) PL approximation f^1 to f. Then a new simple function r^2 is created and deformed into a (finer) approximation f^2 to f, and so on. See figure 1.2 for a single stage of this process when $n = 1$. The deformation can be thought of as a folded piece of paper, and the algorithm follows the path that is the intersection of the

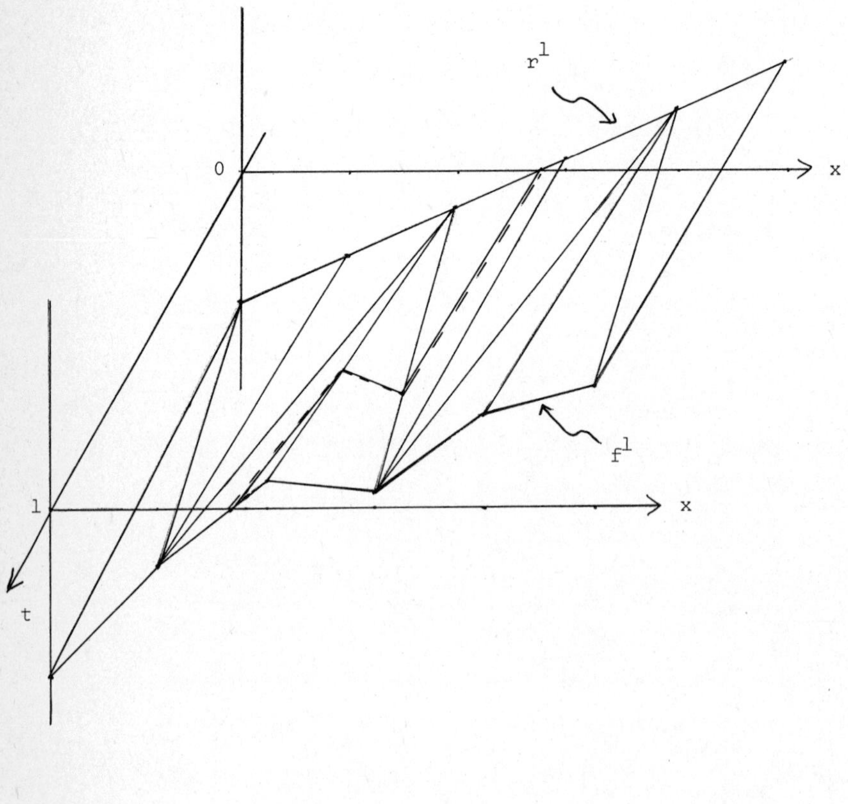

Figure 1.2

paper with a horizontal plane. Clearly, by choosing r^{k+1} appropriately based on the zero x^k of f^k found, this algorithm will generally perform better than the "left-to-right" early methods.

The continuous refinement algorithms link together an infinite sequence of PL approximations as shown in figure 1.3. It can be seen that, in one dimension, this algorithm is merely a geometric realization of the well-known bisection method. However, it is important to note that, for $n > 1$, progress may not be monotonic--for example, after finding a zero of f^k, a second zero of f^{k-1} may be generated before finding a zero of f^{k+1}.

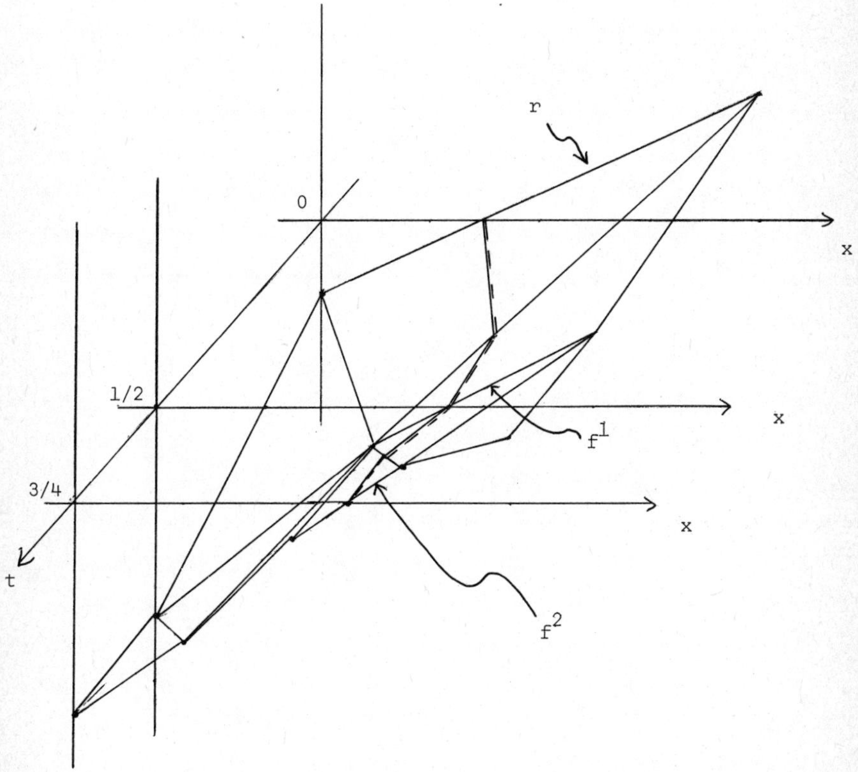

Figure 1.3

A main focus of early research on these methods was to extend their application to unbounded domains, and global convergence conditions were established by a number of authors, including Merrill [55] and Eaves and Saigal [15]. These conditions hold for a variety of important applications, but note that the algorithms can fail on simple problems not suited to them--for example, if $n = 1$, almost all such methods will diverge in attempting to find a zero of $f(x) = x^2$, since almost all piecewise-linear approximations to f have no zero.

In the mid 1970's, there was a considerable amount of work on different triangulations (used in making PL approximations)--see the author's monograph [76] for studies up to 1976, and also [80] and van der Laan and Talman [48]. While work continues in this area, it seems unlikely that one can improve substantially on the simple triangulations now being used.

A significant advance was the work of Saigal [64], see also [67], showing that the algorithms could be implemented to converge quadratically on smooth regular problems. Recently, van der Laan and Talman [45,46,51] have developed several algorithms that save a great deal of the work associated with the extra (homotopy) dimension--see also Tuy, Thoai and Muu [93] and Reiser [60]. A unifying viewpoint for these "variable dimension algorithms" is given in Kojima and Yamamoto [39,40] and Freund [21].

On a more theoretical level, several theorems of topology (e.g., those of Brouwer, Kakutani, Browder, Poincare and Borsuk-Ulam) have been given constructive proofs. See Scarf [69], Eaves [12], Freidenfelds [20], Garcia [24], Meyerson and Wright [56], Bárány [4], Freund and Todd [22] and Todd and Wright [92]. Other existence results have also been established, e.g., Fisher, Gould and Tolle [18], Saigal [63], Gould and Tolle [30], and Charnes, Garcia and Lemke [7]. Great insight into the algorithms has been obtained via index and degree theory, see Eaves and Scarf [16] and Eaves [14].

Some recent research has been concerned with the expoitation of structure in the function f (e.g., separability, sparsity) and to some extent in mappings F. See Kojima [35,36], Todd [84,85] and Awoniyi and Todd [3].

Finally, for certain problems it is possible to obtain not merely one but all solutions. For this it is necessary to know in advance how many solutions there are, so the technique is most useful for complex polynomials or systems of polynomial or polynomial-like equations. See Kuhn [43], Kojima et al. [38], Drexler [11] and Garcia and Zangwill [26,27].

This paper is far from a complete survey of the field. For further reading, we recommend the book of Scarf with Hansen [71], the monographs by Luthi [54] and Todd [76], the short course of Eaves [14] and the review paper of Allgower and Georg [2]. Also of interest are the proceedings edited by Karamardian [31], Peitgen and Walther [59], Forster [19] and Robinson [61]. In particular, the excellent article [2] contains a discussion of the very closely related smooth continuation methods, which we will not discuss in detail here. These originated with the paper of Kellogg, Li and Yorke [34]; see also Smale [74], Alexander and Yorke [1], Keller [33], Chow, Mallet-Paret and Yorke [8] and the book edited by Wacker [95].

The remainder of the paper is organized as follows. In section 2 we discuss the problems to which piecewise-linear homotopy algorithms are applicable and give examples. Section 3 introduces the concepts of triangulation and piecewise-linear approximations. In section 4 we describe some early algorithms. The deficiencies of these methods lead us to homotopies in section 5, where we also briefly describe smooth continuation methods. Section 6 contains descriptions of the restart and continuous deformation algorithms.

Next, we describe some particularly useful triangulations for piecewise-linear homotopy algorithms in section 7. Section 8 outlines an "artificial cube" algorithm, which is one of van der Laan and Talman's variable dimension algorithms. In section 9 we describe the acceleration techniques that can be used for smooth problems. Some recent work on exploiting structure is discussed in section 10. We conclude in section 11 by making a few remarks on computational experience.

Our notation is standard. We use $\|\cdot\|$ for the usual Euclidean norm on vectors and the corresponding operator norm on matrices. Subscripts are generally used for coordinates and superscripts for indexing. For scalars, subscripts are used for indexing and superscripts for powers. There should be no confusion.

2. Problems

Here we describe very briefly the problems we seek to solve and where they arise.

A. Given a continuous function $f: R^n \to R^n$, find x^* with

$$0 = f(x^*). \tag{2.1}$$

If g maps R^n continuously into a compact subset of R^n, we may set $f(x) = x-g(x)$ so that solutions to (2.1) are fixed points of g.

i) Unconstrained Minimization. If $\theta: R^n \to R$ is continuously differentiable we can set $f = \nabla\theta$. Then solutions to (2.1) are stationary points of θ. Note that this approach ignores values of θ and seems as likely to find maxima or saddle points as minima. However, index theory (see, e.g., Todd [78]) implies that if θ is twice continuously differentiable, only points x^* with $\nabla\theta(x^*) = 0$ and $\det \nabla^2\theta(x^*) \geq 0$ may be approximated.

ii) Equilibria in Exchange Economies. Here there is no production: each economic agent enters the marketplace with an initial endowment and after trading with other agents ends with a final allocation of each of $n+1$ goods. At issue is the existence and computation of a price vector p such that, if each agent sells his initial endorsement at prices p and then buys the most preferred bundle of goods he can afford without any considerations of availability, all markets in fact clear. We consider prices in the simplex $S = \{p \in R^{n+1}: p \geq 0$ and $\sum_j p_j = 1\}$, and suppose that a continuous function $z: S \to R^{n+1}$ exists with $z_j(p)$ the excess demand for good j at prices p. This function must satisfy Walras' law $p^T z(p) \leq 0$ for all p from budget considerations. We seek p^* with $z(p^*) \leq 0$, i.e. all markets clear. Define $g: S \to S$ by

$$g_j(p) = \frac{(p_j + z_j(p))_+}{\sum_i (p_i + z_i(p))_+}$$

where λ_+ denotes $\max\{0,\lambda\}$. Then g is continuous and its fixed points are equilibrium price vectors p^* with $z(p^*) \leq 0$. Such a function g can be extended to $\hat{g}$: aff $S \to S$ (where aff S is the affine hull of S) and hence $\tilde{g}: R^n \to \tilde{S} \subseteq R^n$

obtained whose fixed points correspond via a natural projection to equilibria. For more details, see Scarf with Hansen [71], Todd [76], Scarf [70] and Todd [88]. In particular, Sonnenschein and others have shown that virtually any continuous z satisfying Walras' law can arise from a "reasonable" economy--thus the generality of fixed-point theorems is necessary to establish the existence of equilibria--see Scarf [70].

B. Given an upper semi-continuous mapping $F: R^n \to R^{n*}$, where R^{n*} denotes the collection of nonempty compact convex subsets of R^n, find $x*$ with

$$0 \in F(x*). \tag{2.2}$$

i) Unconstrained Minimization. Suppose $\theta: R^n \to R$ is convex, but possibly nonsmooth. Then its subdifferential map, $\partial\theta(x) = \{z \in R^n: \theta(y) \geq \theta(x) + z^T(y-x)$ for all $y \in R^n\}$ maps R^n to R^{n*} upper semi-continuously. If $F = \partial\theta$, solutions to (2.2) minimize θ.

ii) Constrained Minimization. Let $\theta: R^n \to R$ and $g: R^n \to R^m$ be continuously differentiable, and consider the problem of minimizing $\theta(x)$ subject to $g(x) \leq 0$. Define $X_0 = \{x \in R^n: g(x) \leq 0\}$, the feasible region, and $X_i = \{x \in R^n: g_i(x) = \max_j g_j(x) \geq 0\}$ for $i = 1,2,\ldots,m$; X_i is the set of points where constraint i is most violated. Define $f^0 = \nabla\theta$ and $f^i = \nabla g_i$, $i = 1,2,\ldots,m$. Set $F(x)$ to be the convex hull of $\{f^i(x): i \in \{0,1,\ldots,m\}$ and $x \in X_i\}$. Then if $0 \in F(x*)$ and $x* \in X_0$, $x*$ satisfies the necessary conditions of F. John to be a local minimizer of the constrained problem, see, e.g., Awoniyi and Todd [3].

It is easy to extend this example for the case where θ and each g_i is convex and possibly nonsmooth--we have chosen the continuously differentiable formulation to illustrate the exploitation of special structure in mappings F--see section 10.

iii) Economic Equilibrium with Production. See, e.g., Todd [76], Awoniyi and Todd [3].

C. Given a continuous function $h: R^n \times R \to R^n$, and a zero (x^0,λ^0) of h, trace a path of zeroes of h including (x^0,λ^0).

i) Nonlinear Eigenvalue Problem. Consider a simple two-point boundary value problem:

$$-u'' + \lambda\phi(u) = 0, \quad u(0) = u(1) = 0$$

where $\lambda \in R$ is a parameter and $\phi\colon R \to R$ a given nonlinearity with $\phi(0) = 0$. Then, after making a finite difference approximation to $-u''$ we obtain

$$h(x,\lambda) = Ax + \lambda g(x) = 0,$$

where $x = (x_i) \in R^n$, $x_i \approx u(i/(n+1))$, $g(x) = (\phi(x_1),\phi(x_2),\ldots,\phi(x_n))^T/(n+1)^2$ and

$$A = \begin{bmatrix} 2 & -1 & & 0 \\ -1 & 2 & \ddots & \\ & \ddots & \ddots & -1 \\ 0 & & -1 & 2 \end{bmatrix}$$

We wish to examine paths of solutions branching off from the trivial solution $u = 0$ (or $x = 0$). Such paths are often illustrated in a bifurcation diagram, as in figure 2.1. I will not discuss C. further--see, for instance, Peitgen and Walther [59].

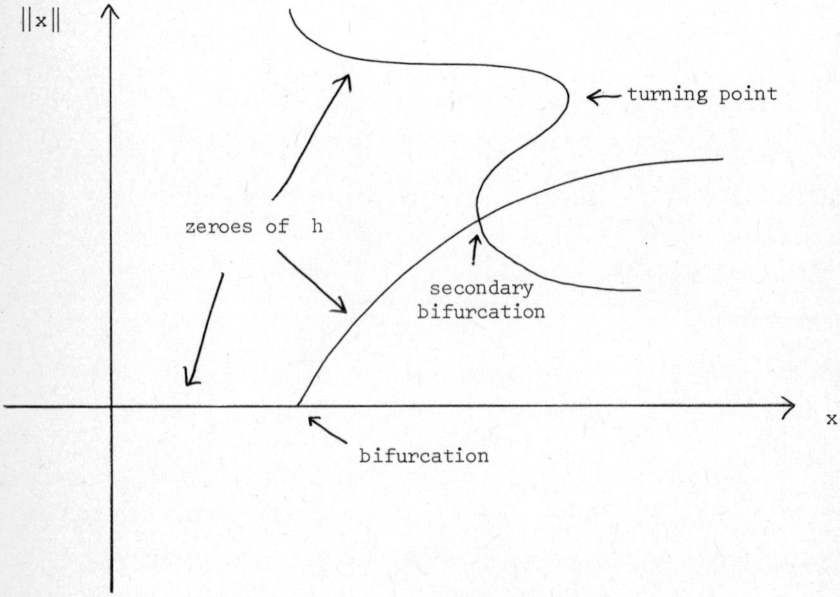

Figure 2.1

3. Triangulations and Piecewise-Linear Approximations

Here we generalize the idea of picking a mesh in R^1 and obtaining corresponding piecewise-linear approximations as in figure 1.1. We also bound the accuracy of these approximations on compact sets.

Definition 3.1. An m-simplex in R^n is the convex hull of m+1 affine independent points, called its vertices. If these are $v^0,v^1,\ldots,v^m$, affine independence means the $(n+1)\times(m+1)$ matrix

$$\begin{pmatrix} 1 & 1 & \ldots & 1 \\ v^0 & v^1 & & v^m \end{pmatrix}$$

has rank m+1. We denote the simplex by $[v^0,v^1,\ldots,v^m]$ and use Greek letters ρ, σ, τ, etc. If the vertices of a j-simplex ρ are all vertices of σ, we say ρ is a j-face (or just face) of σ. The empty set $\emptyset$ is a (-1)-face of every simplex. An (m-1)-face of an m-simplex σ is called a facet of σ.

Definition 3.2. Let $C \subseteq R^n$ be a convex set of dimension m. A collection T of m-simplices in R^n is called a triangulation (or simplicial subdivision) of C iff

a) $\cup\{\sigma: \sigma \in T\} = C$;

b) $\sigma,\tau \in T$ imply $\sigma \cap \tau$ is a face of both σ and τ;

c) (local finiteness) each $x \in C$ has a neighborhood N meeting only finitely many simplices in T.

We let T^k denote the set of k-faces of simplices of T. If v is a vertex of a simplex of T we say v is a vertex of T and write $v \in T^0$ (somewhat abusing the notation). Any $\tau \in T^{m-1}$ is called a facet of T. Let T^+ denote $\cup\{T^k: 0 \leq k \leq m\}$.

Hence $\{[ih,(i+1)h]: 0 \leq i < k\}$ is a triangulation of $[0, kh] \subseteq R^1$, but $\{[-1,0]\} \cup \{[2^{-k-1},2^{-k}]: k = 0,1,\ldots\}$ is not a triangulation of $[-1,1] \subseteq R^1$, since local finiteness is violated. Similarly, figure 3.1(a) is a triangulation of $S^2 = \{x \in R^3: x \geq 0, \sum x_j = 1\}$ while figure 3.1(b) is not.

We will give examples of specific triangulations in section 7. For now we merely give without proof the properties that we use (see, e.g., Todd [76]).

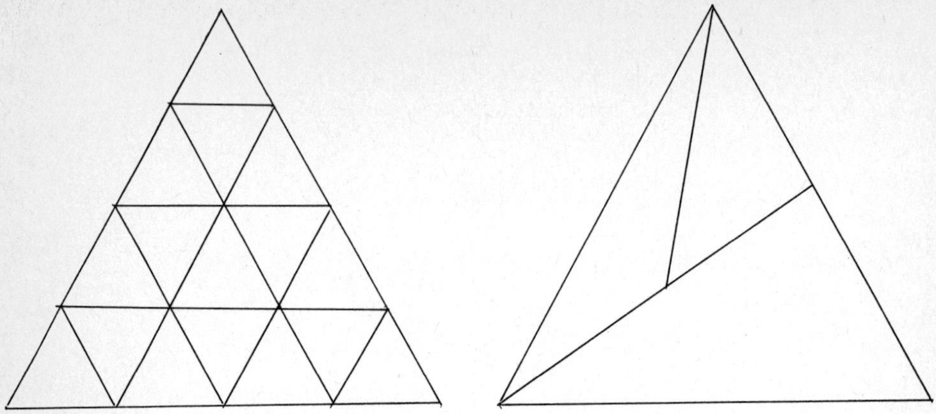

Figure 3.1

Theorem 3.1. Let T be a triangulation of C.

a) If τ is a facet of T, then either

i) $\tau \not\subseteq \partial C$ and τ is a facet of exactly two simplices of T; or

ii) $\tau \subseteq \partial C$ and τ is a facet of precisely one simplex of T.

b) If D is a j-face of C (i.e. $D = H \cap C$ for a hyperplane H where C lies in one of the half-spaces associated with H, and D has dimension j), then $T|_D = \{\tau \in T^j: \tau \subseteq D\}$ is a triangulation of D.

c) If $E \subseteq C$ is compact, $|\{\sigma \in T: \ \sigma \text{ meets } E\}|$ is finite.

Now we consider piecewise-linear (PL) approximations. Let T be a triangulation of $C \subseteq R^n$.

Definition 3.3. Let $f: C \to R^p$ be given. Then the PL approximation to f with respect to T, denoted f_T, is the unique function that agrees with f on the vertices of T and is linear (i.e., affine) on each simplex of T. Thus for any $x \in C$, find $\sigma = [v^0,\ldots,v^m] \in T$ containing x and hence $\lambda_0 \ldots \lambda_m$ with $x = \sum_i \lambda_i v^i$, $\sum_i \lambda_i = 1$, $\lambda_i \geq 0$. Then $f_T(x) = \sum_i \lambda_i f(v^i)$.

Let $F: C \to R^{p^*}$ be given. Then a PL approximation to F, denoted F_T, is a function that is a selection from F at vertices of T and is linear on each simplex of T. Thus $F_T(v) \in F(v)$ for each $v \in T^0$ and $F_T(\sum_i \lambda_i v^i) = \sum_i \lambda_i F_T(v^i)$ whenever $[v^0,\ldots,v^m] \in T$, $\sum \lambda_i = 1$ and $\lambda_i \geq 0$.

It is easy to see (using the local finiteness of T) that f_T and F_T are continuous functions. In order to describe the degree to which they approximate f and F, we must introduce a notion of the "fineness" of T.

Definition 3.4. The mesh size of T, denoted mesh T, is defined to be $\sup\{\text{diam } \sigma: \sigma \in T\}$, where $\text{diam } \sigma$ denotes the diameter $\max\{\|x-y\|: x,y \in \sigma\}$ of σ.

Proposition 3.1. Let $f: C \to R^p$ be uniformly continuous with modulus of continuity ω and let T be a triangulation of C of mesh size at most δ. Then $\|f_T - f\|_C \equiv \sup\{\|f_T(x) - f(x)\| : x \in C\} \leq \omega(\delta)$. Hence $f_T(x^*) = 0$ implies $\|f(x^*)\| \leq \omega(\delta)$.

Proof. Let $x \in C$ be arbitrary and write $x = \sum_i \lambda_i v^i$, $[v^0,\ldots,v^m] \in T$, $\sum_i \lambda_i = 1$, $\lambda_i \geq 0$. Then $\|f(v^i)-f(x)\| \leq \omega(\delta)$, so $\|\sum_i \lambda_i f(v^i) - f(x)\| \leq \omega(\delta)$.

The approximation is better if f is smooth.

Proposition 3.2. Let $f: C \to R^p$ be continuously differentiable on C, and suppose its derivative matrix Df satisfies a Lipschitz condition with constant κ on C:

$$\|Df(x) - Df(y)\| \leq \kappa \|x-y\|, \quad x,y \in C. \tag{3.1}$$

Then, if T is a triangulation of C of mesh size at most δ, $\|f_T - f\|_C \leq \frac{1}{2}\kappa\delta^2$.

Proof. We use the following lemma, e.g. Ortega and Rheinboldt [58], p. 73: given condition (3.1),

$$\|f(w) - f(z) - Df(z)(w-z)\| \leq \frac{1}{2}\kappa \|w-z\|^2 \tag{3.2}$$

for all w,z in C. Hence, with x, v^i's and λ_i's as above,

$$f(v^i) - f(x) - Df(x)(v^i - x) = r^i, \quad \|r^i\| \leq \frac{1}{2}\kappa\delta^2,$$

and $\sum_i \lambda_i f(v^i) - f(x) - Df(x)(\sum_i \lambda_i v^i - x) = \sum_i \lambda_i r^i$, i.e.

$$\|f_T(x)-f(x)\| \leq \frac{1}{2}\kappa\delta^2 \quad \text{as required.}$$

Finally, we state a result appropriate to the point-to-set case. For details, see Scarf with Hansen [71]. (Note that Todd [76] claims such an approximation result, but his hypotheses do not follow from upper semi-continuity.) Here $B(y,\eta)$ denotes $\{z: \|z-y\| \leq \eta\}$.

Proposition 3.3. Let $F: C \to R^{p^*}$ be upper semi-continuous, where C is compact. Then for each $\varepsilon > 0$ there exists $\delta > 0$ such that, if T is a triangulation of C of mesh size at most δ, then for all $x \in C$, $F_T(x) \in B(F(B(x,\varepsilon)),\varepsilon)$.

Thus if $F_T(x) = 0$, x is close to a point whose image, when expanded a little, contains 0.

The proposition can be established in a straightforward manner from the following lemma, whose proof (by contradiction) is omitted.

Lemma 3.1. For all $\varepsilon > 0$ there exists $\delta > 0$ such that $\bar{x} \in C$ and $(\bar{x},\bar{f}) \in$ $\mathrm{conv}(\{(x,f): x \in C, f \in F(x)\} \cap (B(\bar{x},\delta)\times R^n))$ imply $\bar{f} \in B(F(B(\bar{x},\varepsilon)),\varepsilon)$.

4. Early Algorithms

This section outlines briefly a typical example of an early "fixed-grid" algorithm, and illustrates it in dimension $n = 2$. The deficiencies of this algorithm will motivate the introduction of a homotopy in subsequent sections.

What follows is a description of the "variable dimension" algorithm of Kuhn [42] and Shapley [73] for approximating a fixed point of a continuous function $g: S^n \to S^n$, where S^n is the n-dimensional simplex $\{x = (x_0,x_1,\ldots,x_n)^T \in R^{n+1}: x \geq 0, \sum_j x_j = 1\}$. To see how this algorithm yields constructive proofs of Sperner's lemma and Brouwer's fixed-point theorem, see e.g. [76].

Let T be a triangulation of S^n of mesh size at most δ. Instead of finding a fixed point of the PL approximation g_T to g, we shall assign to each $v \in T^0$ an integer label $i = L(v) \in \{0,1,\ldots,n\}$ such that $g_i(v) \leq v_i$ and $v_i > 0$. Such an index i always exists, since both the $g_j(v)$'s and the v_j's sum to one. We may choose $L(v)$ to be the smallest such index, for example. We will search for a simplex $\sigma = [v^0,\dots,v^n] \in T$ such that $L(v^i) = i$, $i = 0,1,\dots,n$. Such a simplex is called completely labelled. In order to relate this to our work in the previous section, we now define a (terribly coarse) approximation $\tilde{g}_T$ to g.

Let $\gamma = \min\{\min\{v_i: i \text{ s.t. } v_i > 0\}: v \in T^0\}$. Then for each $v \in T^0$ we may set, with $L(v) = i$,

$$\tilde{g}_T(v) = v + (\gamma/n)(e - (n+1)e^i)$$

where e, here and below, denotes a vector of ones of appropriate dimension and e^i the ith unit vector of appropriate dimension (in this case, with $i \in \{0,1,\dots,n\}$). We next extend $\tilde{g}_T$ linearly on each simplex of T. It is clear that $\tilde{g}_T$ is far from a reasonable approximation to g, although it shares with g the property that $(\tilde{g}_T(v))_i \le v_i$ if $v \in T^0$ and $L(v) = i$. Nevertheless, it is easy to prove (see [71,76])

Proposition 4.1. Suppose g has modulus of continuity ω on S^n. Then if $\tilde{g}_T(x) = x$, $\|g(x)-x\| \le \sqrt{n^2+n}\,(\delta+\omega(\delta))$.

Note that $\tilde{g}_T(x) = x$ iff x is the barycenter of a completely labelled simplex of T. We now describe how such a completely labelled simplex can be found.

Definition 4.1. A simplex $\sigma = [v^0,\dots,v^m] \in T^+$ is completely labelled (c.l.) [respectively, almost completely labelled (a.c.l.)] if $\{L(v^i): i = 0,\dots,m\} \supseteq \{0,1,\dots,n\}$ [respectively, $\{0,1,\dots,n-1\}$]. The graph G^n has as nodes all a.c.l. simplices of T^n and all a.c.l. simplices of T^{n-1} that lie in ∂S^n (and, since they are a.c.l., thus in $\hat{S}^{n-1} \equiv \{x \in S^n: x_n = 0\}$). Two nodes of G^n are adjacent if their intersection is a.c.l.

Figure 4.1 illustrates a labelled triangulation for $n = 2$, and the graph G^2 is shown.

It is easy to verify (using theorem 3.1) that an a.c.l. facet of T in $\hat{S}^{n-1}$ or a c.l. simplex of T has degree one in G^n, while all other nodes have degree two. (The use of such arguments to establish convergence of mathematical programming algorithms was introduced by Lemke and Howson [53].) This yields the following result, illustrated in figure 4.1:

Theorem 4.1. Each connected component of G^n is a cycle or a path each of whose endpoints is an almost completely labelled facet in $\hat{S}^{n-1}$ or a completely labelled simplex.

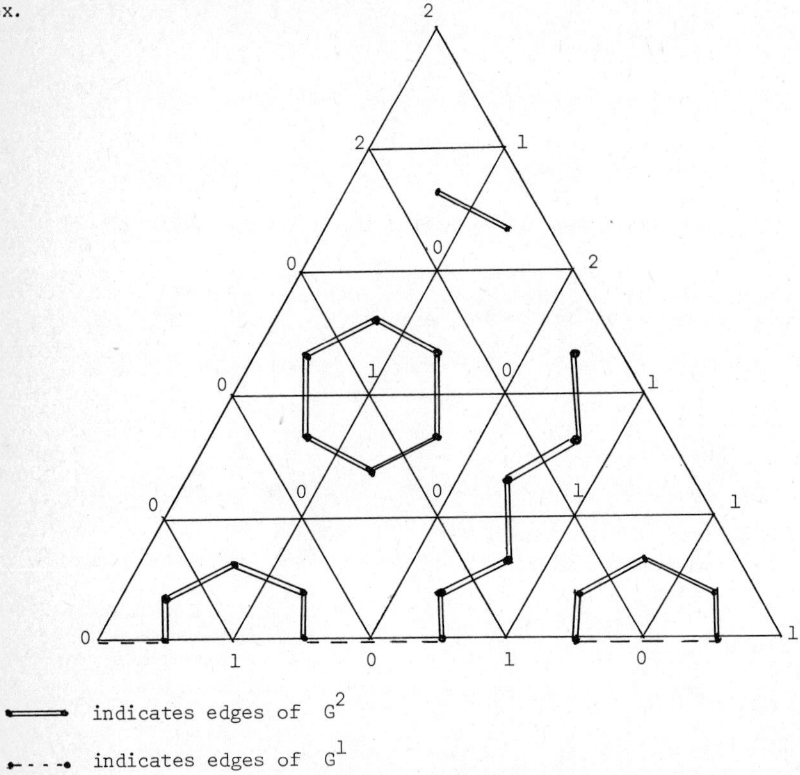

Figure 4.1

Now there is a natural identification of $\hat{S}^{n-1}$ with S^{n-1}, obtained by dropping the final (zero) coordinate, and theorem 3.1 shows that $\hat{T} = T|_{\hat{S}^{n-1}}$ is a triangulation of $\hat{S}^{n-1}$. Also, a.c.l. facets of T in $\hat{S}^{n-1}$ can be viewed as completely labelled simplices of $\hat{T}$. Since the number of endpoints of the paths in G^n is even, the parities of completely labelled simplices in T and in $\hat{T}$ are equal. Thus we have an inductive proof of the strong form of Sperner's lemma, that there is an odd number of completely labelled simplices in T. This semi-constructive proof is due to Cohen [9]. (The basis, for $n = 0$, is trivial.)

To obtain a constructive proof, we define graphs G^j, $1 \leq j \leq n$. Nodes of

G^j are simplices of T^j in $\hat{S}^j \equiv \{x \in S^n: x_k = 0$ for $k > j\}$ and simplices of T^{j-1} in $\hat{S}^{j-1}$, whose vertices have the labels $0,1,\ldots,j-1$ at least. Two such are adjacent if their intersection has all these labels. G^1 is also illustrated in figure 4.1. From theorem 4.1 (extended to all j) we obtain

Theorem 4.2. Each connected component of $G^+ = \bigcup_{j=1}^{n} G^j$ is a cycle, a path whose endpoints are two completely labelled simplices of T, or a path joining the 0-simplex $[e^0]$ with a completely labelled simplex of T.

Thus the algorithm of Kuhn and Shapley is to trace this last path. We start with the 0-simplex $[e^0]$ with label 0, search the edge $\hat{S}^1$ to find a 1-simplex $[v^0,v^1]$ with labels 0,1, then the triangle $\hat{S}^2$ to find a 2-simplex with labels 0,1,2, and so on. We may be forced to return to a lower-dimensional face as in the example of figure 4.1, but in a finite number of steps we must find a completely labelled simplex. With standard triangulations, finding the successive simplices of varying dimension is easy. Obviously, we only evaluate $g(v)$ (to obtain $L(v)$) for a vertex $v \in T^0$ when it is encountered by the path we are tracing. Thus we might hope that only a very small fraction of the number of simplices (and of their vertices) will be generated.

From the preceding description, the drawbacks of this algorithm are fairly apparent. Note first that $\tilde{g}_T$ is a very poor approximation to g. While proposition 4.1 gives some hope of obtaining approximate solutions, it seems that for a given accuracy the mesh size δ will need to be much smaller than when using PL approximations, at least when g is smooth--compare propositions 4.1 and 3.2. Next, the algorithm starts at a vertex of S^n. By relabelling, any vertex can be chosen, but, if δ is small, each such choice is likely to lead to a large number of iterations. Moreover, even if a fixed point is suspected to lie in a particular region of S^n, this information cannot be exploited, except by choosing the closest vertex.

There are several other early algorithms. We have chosen not to describe the original Scarf algorithm, since Scarf uses the related notion of primitive sets rather than simplices of triangulations, on which most recent methods are based.

Also, Kuhn [41] has an algorithm allowing starts anywhere on the boundary of S^n, and Scarf and Hansen [71] and Eaves [12] developed algorithms that use PL approximations g_T rather than approximations $\tilde{g}_T$. Since $v-g_T(v)$ can be any vector, in contrast to the finite number $(n+1)$ of possibilities for $v-\tilde{g}_T(v)$, algorithms using PL approximations g_T are often called vector-labelling, whereas those using functions of the form $\tilde{g}_T$, or, equivalently, the mapping $L: T^0 \to \{0,1,\ldots,n\}$, are termed integer-labelling. Such vector-labelling methods can handle point-to-set mappings in addition to continuous functions. However, all these algorithms share the property that they must start on the boundary of the domain of interest and thus cannot exploit information concerning the location of a solution. In particular, it is therefore costly to obtain very accurate solutions with such methods.

5. Homotopies

To avoid the computational inefficiency of the early algorithms, we introduce the concept of a homotopy. This is a one-parameter family of functions including the function of interest. The increased dimension gives the flexibility to start the algorithm wherever we wish, rather than on the boundary of the region of interest. Moreover, the extra dimension automatically generates paths of solutions, without having to introduce some freedom as in almost completely labelled simplices. (However, it is possible to interpret the early algorithms in a homotopy setting--see Eaves [14] and Eaves and Scarf [16].)

Let us assume that we seek a zero of a continuous function $f: R^n \to R^n$. We then construct a homotopy $h: R^n \times [0,1] \to R^n$ with $h(\cdot,1) = f$ and $h(\cdot,0)$ a function r with known zero $x^0 \in R^n$. In particular, we can choose for h the convex homotopy

$$h(x,t) = tf(x) + (1-t)r(x). \tag{5.1}$$

If r is a translated version of f, i.e., $r(x) = f(x)-f(x^0)$, then

$$h(x,t) = f(x) - (1-t)f(x^0) \tag{5.2}$$

and thus all zeroes of h satisfy $f(x) = (1-t)f(x^0)$. This homotopy is frequently called the global Newton homotopy, for reasons that will become clearer below. It is a classical choice for so-called embedding methods: e.g., if $0 = t_0 < t_1 < \ldots < t_k = 1$, we may solve $h(x,t_i) = 0$ for $x = x^i$ using some local method starting at $x = x^{i-1}$, for $i = 1,2,\ldots,k$. For a discussion of such methods, see section 7.5 of Ortega and Rheinboldt [58].

Another simple choice for r is an affine isomorphism, i.e., $r(x) = A(x-x^0)$, so that

$$h(x,t) = tf(x) + (1-t)A(x-x^0) \qquad (5.3)$$

where A is an $n \times n$ nonsingular matrix. We will frequently make such a choice below.

However h is constructed, our hope is that the zeroes of h form a set of "loops" and "paths", one of which starts at the point $(x^0,0)$. See figure 5.1. We aim to follow this path, hoping that it leads to a point $(x^*,1)$; in this case, x^* is a zero of f as desired. This paper is concerned with methods that

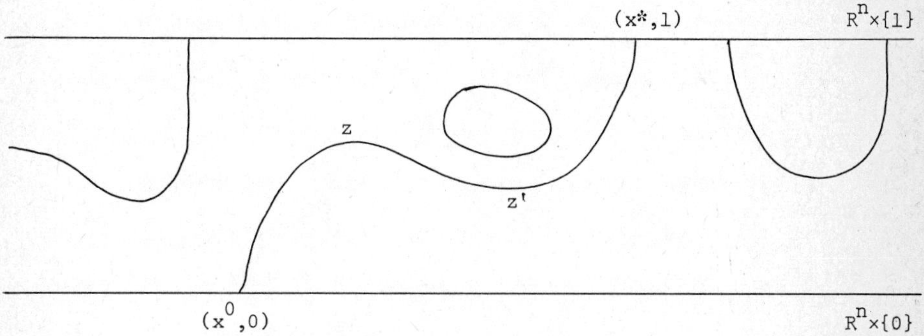

Zeroes of a smooth, regular h

Figure 5.1

approximate this path by making PL approximations to h, and trace the PL path of zeroes of such approximations. Note that if instead we seek a zero of a point-to-set mapping $F: R^n \to R^{n*}$, we may construct a homotopy $H: R^n \times [0,1] \to R^{n*}$, e.g.

$$H(x,t) = tF(x) + (1-t)\{r(x)\} \tag{5.4}$$

and again consider zeroes of PL approximations to H. This possibility demonstrates the generality of the PL approach, but now we wish to consider briefly other methods using the continuous homotopy h of (5.1), assuming f and r are smooth.

First suppose we approximate h in (5.1) globally by truncating its Taylor expansion after the first-order terms. If the expansion is based on the point (x^0, t_0), let this linear approximation be $\hat{h}$. Then the following cases demonstrate that the homotopy approach contains many well-known methods. We assume that all appropriate matrices are nonsingular.

a) h given by (5.2). Then for all t_0, $\hat{h}(\hat{x},1) = 0$ for $\hat{x} = x^0 - (Df(x^0))^{-1} \cdot f(x^0)$, the Newton iterate from x^0.

b) h given by (5.3). For $t_0 = 1$, solving $\hat{h}(\hat{x},1) = 0$ yields $\hat{x}$ as in (a). For $t_0 = 0$, $\hat{h}(\hat{x},1) = 0$ for $\hat{x} = x^0 - A^{-1}f(x^0)$, a Newton-like iterate from x^0; if $f = \nabla\theta$ and $A = \alpha I$, $\hat{x}$ is a steepest descent step from x^0. For $0 < t_0 < 1$, $\hat{h}(\hat{x},1) = 0$ for $\hat{x} = x^0 - (t_0 Df(x^0) + (1-t_0)A)^{-1}f(x^0)$; if $f = \nabla\theta$ and $A = \alpha I$, this corresponds to a step taken in the algorithm of Goldfeld, Quandt and Trotter [29].

Next we discuss methods based on solving initial value problems. First suppose that for all $t \in [0,1]$ there is a solution $x = x(t)$ to $h(x,t) = 0$, and $x(t)$ depends continuously on t with $x(0) = x^0$. Let us further assume that $h_x(x,t)$ (the $n \times n$ matrix that is the derivative of h with respect to x) is nonsingular at each point $(x(t),t)$. Then we obtain

$$\dot{x}(t) = -[h_x(x(t),t)]^{-1} h_t(x(t),t). \tag{5.5}$$

If h is as in (5.2), this becomes $\dot{x} = -(Df(x))^{-1}f(x^0)$ or, for $t < 1$,

$$\dot{x} = -(Df(x))^{-1}f(x)/(1-t). \tag{5.6}$$

Thus we may solve for $x(t)$ by solving the initial value problem (5.6) with $x(0) = x^0$. Note that the corresponding path always moves in the Newton direction; the method is a continuous analogue of Newton's method. However, the algorithm will fail if it finds a point $x = x(t)$ with $Df(x)$ singular: in figure 5.1, z and z' are such points. For a discussion of such methods, see section 7.5 of Ortega and Rheinboldt [58].

In order to traverse points z and z', we parametrize by arc length instead of t. Thus we have the initial value problem

$$\begin{aligned} h_x\dot{x} + h_t\dot{t} &= 0 \\ \|(\dot{x},\dot{t})\| &= 1 \\ x(0) = x^0,\ t(0) &= 0. \end{aligned} \tag{5.7}$$

Note that if h is given by (5.2), then for points (x,t) on the path with $Df(x)$ nonsingular and $t < 1$, we have

$$\begin{aligned} \dot{x} &= -\lambda(Df(x))^{-1}f(x), \\ \lambda &= \dot{t}/(1-t) \end{aligned} \tag{5.8}$$

so that x is moving either in the positive or negative Newton direction. (In figure 5.1, the negative direction is taken between z and z'.) Such a method was considered first by Branin [6]; for a complete discussion see Allgower and Georg [2].

In order that the zeroes of h form paths and loops, it is sufficient by the implicit function theorem that the $n\times(n+1)$ matrix (h_x,h_t) have rank n at all points (x,t) in $h^{-1}(0)$. This regularity condition holds for almost all additive perturbations of h by Sard's theorem, e.g., [2]; alternatively, the "artificial" function r in (5.1) can be chosen "at random"--see Chow, Mallet-Paret and Yorke [8].

To conclude this section we consider PL approximations h_T to h. In general, we will let T be a triangulation of $R^n\times[0,1]$. However, note that if h is given by (5.2), t appears linearly with constant coefficient. Thus, if $\hat{T}$ is a triangulation of R^n, we can set $h_T(x,t) = f_{\hat{T}}(x) - (1-t)f(x^0)$; h_T is piecewise-linear, with pieces of the form $\sigma\times[0,1]$, $\sigma \in \hat{T}$, giving a polyhedral subdivision T of $R^n\times[0,1]$. This algorithm is basically that of Katzenelson [32], Eaves [12] and Garcia [23]. Note that by replacing $f(x^0)$ by $f_{\hat{T}}(x^0)$, $(x^0,0)$ is a zero of h_T. We will also have occasion to use special triangulations T of $R^n\times[0,1)$, so that h_T is defined only on $R^n\times[0,1)$. However, by defining $h_T(\cdot,1) = f$, we will obtain a continuous function h_T: $R^n\times[0,1] \to R^n$.

Suppose now T is a triangulation of $R^n\times[0,1]$. Then by projecting $T|_{R^n\times\{1\}}$ onto R^n, we obtain a triangulation $T(1)$ of R^n. Note that a zero of h_T in $R^n\times\{1\}$ yields a zero of $f_{T(1)}$. Moreover, if we choose h as in (5.3) then $h(\cdot,0) = r$ is linear, so that $h_T(\cdot,0) = r$ and h_T has a <u>unique</u> zero $(x^0,0)$ in $R^n\times\{0\}$. Thus in this case, if $h_T^{-1}(0)$ is a collection of loops and paths, the path starting at $(x^0,0)$ either ends at $(x^1,1)$, with x^1 a zero of $f_{T(1)}$, or goes to infinity. Hence if suitable boundary conditions preclude the latter possibility, $f_{T(1)}$ has a zero, and by taking a sequence T_k of triangulations with mesh $T_k(1)$ $\to 0$, we have a constructive proof of existence of a zero of f.

Finally we state without proof some results guaranteeing that $h_T^{-1}(0)$ or $h_T^{-1}(y)$ is a collection of loops and paths. These results also hold if h_T is replaced by H_T.

<u>Definition 5.1</u>. y is a regular value of h_T (restricted to C) if $h_T(x) = y$ (and $x \in C$) imply that x lies in no simplex of T^+ of dimension less than n.

<u>Proposition 5.1</u>. Let $C \subseteq R^n\times[0,1]$ be compact. Then there exists $\bar{\varepsilon} > 0$ such that for all $0 < \varepsilon \leq \bar{\varepsilon}$, $\langle\varepsilon\rangle \equiv (\varepsilon,\varepsilon^2,\ldots,\varepsilon^n)^T$ is a regular value of h_T restricted to C.

This is a PL version of Sard's theorem. See Eaves [14].

<u>Definition 5.2</u>. A regular PL route in T is a connected subset of $R^n\times[0,1]$ whose intersection with each simplex of T is either empty or a line segment joining two points in the relative interiors of two distinct facets of the simplex.

Such a route is homeomorphic to either the circle or an interval. Correspondingly we call it a loop or a path. Each path has 0, 1 or 2 endpoints, which are the images of endpoints of the homeomorphic interval.

<u>Theorem 5.1</u>. If y is a regular value of h_T, then $h_T^{-1}(y)$ is a disjoint union of regular PL loops and paths, and each endpoint of a path lies in $R^n\times\{0\}$ or $R^n\times\{1\}$.

For a proof, see Eaves [14].

6. Restart and Continuous Deformation Algorithms

Here we describe Merrill's restart algorithm [55] and (briefly) Eaves and Saigal's continuous deformation algorithm [15] and give a condition due to Merrill guaranteeing global convergence.

For Merrill's algorithm, we define h by (5.3) and let T be a triangulation of $R^n\times[0,1]$ with $T^0 \subseteq R^n\times\{0,1\}$ and projected mesh size p mesh $T \equiv \sup\{\max\{\|x^1-x^2\| : (x^i,t_i)\in\sigma,\ i=1,2\}: \sigma\in T\}$ finite. We note that $h_T(x,0)=0$ iff $x=x^0$. A <u>major cycle</u> of Merrill's algorithm entails tracing a path in $h_T^{-1}(\langle\varepsilon\rangle)$ for small positive ε from a point near $(x^0,0)$. (Recall $\langle\varepsilon\rangle = (\varepsilon,\varepsilon^2,\ldots,\varepsilon^n)^T$.) We will assume that T is translated if necessary so that $(x^0,0)$ lies in the relative interior of a facet of T; this facet will then contain a point in $h_T^{-1}(\langle\varepsilon\rangle)$ for all sufficiently small positive ε. This path is traced numerically by considering certain linear systems, as we shall see below.

<u>Example 6.1</u>. Let $n=1$, $f(x)=x^3+2$ and $r(x)=x-1/2$. Then figure 6.1 illustrates a triangulation of $R^1\times[0,1]$ and the path $h_T^{-1}(0)$. Next to each vertex v we mark $h(v)$.

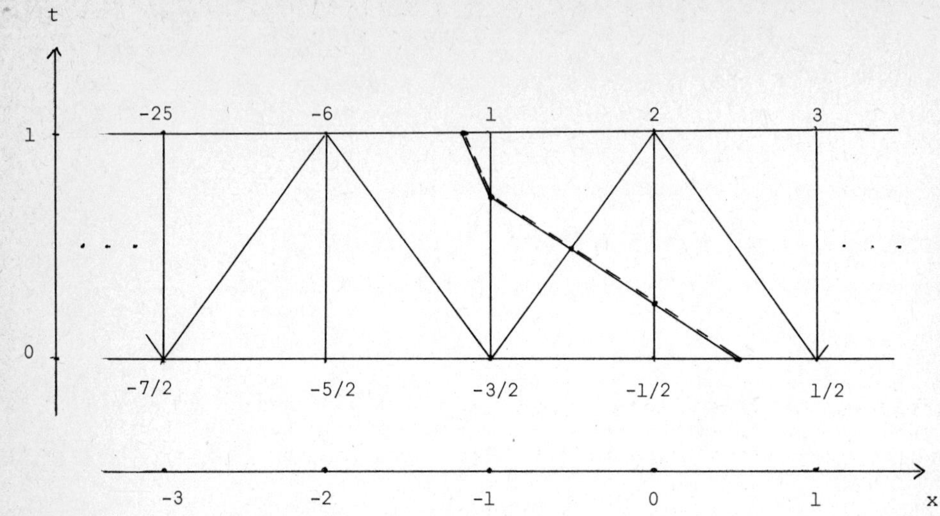

Figure 6.1

Definition 6.1. Let $\sigma = [v^0, v^1, \ldots, v^m]$ be a simplex of $T^n \cup T^{n+1}$. Then the label matrix of σ is the $n \times (m+1)$ matrix

$$L_\sigma = \begin{pmatrix} 1 & \ldots & 1 \\ h(v^0) & & h(v^m) \end{pmatrix}.$$

We say σ is complete (very complete) if there is a feasible solution to

$$L_\sigma w = (1,0,\ldots,0)^T, \quad w \geq 0 \tag{6.1}$$

$$(L_\sigma W = I, \quad W \succeq 0). \tag{6.2}$$

Here $W \succeq 0$ means that each row of W is zero or has first nonzero component positive: we say W is lexicographically nonnegative.

By postmultiplying $L_\sigma W = I$ by $(1, \varepsilon, \varepsilon^2, \ldots, \varepsilon^n)^T$, we obtain

Lemma 6.1. $L_\sigma w = (1,\varepsilon,\varepsilon^2,\ldots,\varepsilon^n)^T$, $w \geq 0$, has a feasible solution for all sufficiently small positive ε iff σ is very complete.

Theorem 6.1. A simplex $\sigma \in T^n \cup T^{n+1}$ meets $h_T^{-1}(0)$ ($h_T^{-1}(\langle\varepsilon\rangle)$ for all sufficiently small positive ε) iff σ is complete (very complete).

Proof. Let $\sigma = [v^0,\ldots,v^m]$ and let $x = \sum_i w_i v^i$, $\sum_i w_i = 1$, $w_i \geq 0$, be in $\sigma \cap h_T^{-1}(0)$. Then w solves (6.1). Conversely, if w solves (6.1) then $x = \sum_i w_i v^i$ lies in σ and $h_T(x) = 0$. The other part follows similarly using lemma 6.1.

The theorem shows that the path in $h_T^{-1}(0)$ (or $h_T^{-1}(\langle\varepsilon\rangle)$) can be followed by considering the linear systems (6.1) (or (6.2)); moreover, we can trace a path in $h_T^{-1}(\langle\varepsilon\rangle)$ for all sufficiently small positive ε by considering the lexicographic system (6.2). This implies that the dependence of $\bar{\varepsilon}$ in proposition 5.1 on C is immaterial: at any point within the major cycle, all simplices generated will lie in some compact set C and an appropriate positive $\bar{\varepsilon}$ exists, but $\bar{\varepsilon}$ need not be known.

If σ is an n-simplex then L_σ is square, and there is usually a unique solution to (6.1) if σ is complete, and definitely a unique solution if σ is very complete. If σ is an (n+1)-simplex, then there is typically some freedom in (6.1) or (6.2) since L_σ has one more column than row. Indeed, if σ is very complete, then L_σ has rank $n+1$ and all solutions to (6.2) are of the form $W = \bar{W} + du^T$, $u \in R^{n+2}$, where d is in the null space of L_σ. Since $e^T W = (1,0,\ldots,0)$, it is not hard to show that precisely two such W's have a row of zeroes. See, e.g., Dantzig [10]. Hence we obtain

Proposition 6.1. If τ is a facet of $\sigma \in T^{n+1}$, then σ is complete (very complete) if τ is. If $\sigma \in T^{n+1}$ is very complete, σ has precisely two very complete facets.

From the nonsingularity of A in (5.3) and the fact that T was chosen so that $(x^0,0)$ was in the relative interior of a facet, we get

Proposition 6.2. If τ_0 is the facet of T containing $(x^0,0)$, τ_0 is very complete.

We can now state how one major cycle of Merrill's algorithm is performed.

Step 0. Let σ_0 be the unique simplex of T with τ_0 as a facet. Set $p \leftarrow 0$.

Step 1. τ_p is a very complete facet of σ_p. Let τ_{p+1} be its other very complete facet.

Step 2. If $\tau_{p+1} \subseteq R^n \times \{1\}$ go to step 3. Otherwise, let $\sigma_{p+1} \neq \sigma_p$ be a simplex of T with τ_{p+1} as a facet. Set $p \leftarrow p+1$ and go to step 1.

Step 3. $\tau_{p+1} = [v^0,\ldots,v^n] = [(y^0,1),\ldots,(y^n,1)]$ is very complete, hence complete. Thus there exist weights $w_0 \ldots w_n$ with $\bar{z} = \sum_i w_i v^i$ a zero of h_T and $\bar{x} = \sum_i w_i y^i$ a zero of $f_{T(1)}$. Stop.

Note that we only need to evaluate h at vertices v that arise as the vertex of σ_p that is not a vertex of τ_p in step 1. The vector $h(v)$ then gives a new column of L_{σ_p}. Obtaining τ_{p+1} then amounts to a (lexicographic) linear programming pivot step in the linear system (6.2). In step 2, the simplex σ_{p+1} is of course unique, see theorem 3.1. The triangulations we shall describe in section 7 allow very simple generation of σ_{p+1} given σ_p and τ_{p+1}.

The Lemke-Howson argument [53] alluded to in section 4 shows that the algorithm will produce a sequence $\tau_0,\sigma_0,\tau_1,\ldots$ of distinct n- and $(n+1)$-simplices, which must therefore either terminate in step 3 or diverge (theorem 3.1(c)). Before investigating how such divergence can be precluded, let us describe how major cycles are linked together in such restart algorithms.

If the major cycle terminates in step 3 with $\bar{x}$, let us write $\bar{x} = M(f,x^0,A,T)$. Let $\{T_k\}_{k=1}^{\infty}$ be a sequence of triangulations of $R^n \times [0,1]$, with $T_k^0 \subseteq R^n \times \{0,1\}$ and p mesh $T_k \to 0$. Then, assuming that each T_k is such that $(x^{k-1},0)$ lies in the relative interior of a facet of T_k, the overall algorithm is defined by

$$x^k = M(f,x^{k-1},A,T_k), \quad k = 1,2,\ldots, \tag{6.3}$$

if all major cycles terminate.

If we seek a zero of a mapping $F\colon R^n \to R^{n*}$, we construct H using (5.4) with $r(x) = A(x-x^0)$ and hence H_T. With appropriate changes to the preceding, we may then write $\bar{x} = M(F,x^0,A,T)$ if the major cycle terminates; in this case $\bar{x}$ is a zero of some $F_{T(1)}$. The overall algorithm is then as in (6.3) with F replacing f.

Suppose the sequence $\{x^k\}$ defined by (6.3) exists and x^* is a limit point. Then proposition 3.1 implies that x^* is a zero of f. Now replace f by F, an upper semi-continuous mapping from R^n to R^{n*}. Then if x^* is a limit point, $0 \in F(x^*)$, since otherwise there is a half-space $S = \{y: d^T y > \gamma\}$ with $\gamma > 0$ containing $F(x^*)$, and, by upper semi-continuity, $F(x)$ for all x close to x^*. Then proposition 3.3 gives a contradiction.

We now give a condition ensuring that all major cycles terminate. This is based on the work of Merrill [55], who used $A = B = -I$ in the following.

<u>Condition 6.1</u>. Suppose $f: R^n \to R^n$ satisfies: For all $\delta > 0$, there exist $c \in R^n$, $C \subseteq R^n$ bounded, and nonsingular $n \times n$ matrix B with $A^T B$ positive definite (not necessarily symmetric), and $y \notin C$, $\|x-y\| \leq \delta$ imply $f(y)^T B(x-c) > 0$.

For a point-to-set mapping F, the final inequality is replaced by "$f^T B(x-c) > 0$ for all $f \in F(y)$."

Note that, if $A = B = I$ and $c = 0$ then condition 6.1 is a slight strengthening of the requirement that $f(x)$ point "outward" for large $\|x\|$. The introduction of $B \neq A$ allows A to incorporate for instance different scales for the variables, while B remains $\pm I$.

We state the following theorem for a continuous function--it holds with appropriate changes also for point-to-set mappings.

<u>Theorem 6.2</u>. Let f satisfy condition 6.1 and define $\{x^k\}$ by (6.3) where x^0 is arbitrary and each T_k has $T_k^0 \subseteq R^n \times [0,1]$ and $(x^{k-1},0)$ in the relative interior of a facet. Let p mesh $T_k \to 0$. Then $\{x^k\}$ is well-defined and bounded. Hence it has a limit point x^* and any such limit point is a zero of f.

<u>Proof</u>. The last part follows from the remarks above. To show that $\{x^k\}$ is well-defined and bounded, let $\delta = \max_k p \text{ mesh } T_k$, and let c, C and B correspond

by condition 6.1. Without loss of generality, assume $x^0 \in C$. If x^k exists for $k \geq 1$, it is a zero of $f_{T_k(1)}$, where $T_k(1)$ has mesh size at most δ. We show first that all such x^k's lie in the bounded set $C' = \{x \in R^n: \|x-y\| \leq \delta$ for some $y \in C\}$. Indeed, let $x = \sum_i w_i y^i$ with $\sum_i w_i = 1$ and $w_i \geq 0$, where $[y^0,\ldots,y^n] \in T_k(1)$. If $x \notin C'$, $y^i \notin C$, so that $f(y^i)^T B(x-c) > 0$ by condition 6.1. Hence $f_{T_k(1)}(x)^T B(x-c) > 0$, and thus $f_{T_k(1)}(x) \neq 0$. Thus all x^k's lie in C', since x^0 clearly does.

Now we must show that all major cycles converge. Suppose the first $k-1$ do, so that $x^0, x^1, \ldots, x^{k-1}$ exist and lie in C'. We show that all complete simplices generated in the kth major cycle lie in a bounded set $D' \times [0,1]$. Since this contains only finitely many simplices of T_k, the proof will be complete.

Let $D \subseteq R^n$ be a bounded set containing C' and with the property that $y \notin D$ and $\|x-y\| \leq \delta$ imply $(y-x^{k-1})^T A^T B(x-c) > 0$. Such a set exists because

$$(y-x^{k-1})^T A^T B(x-c) \geq \lambda \|y-x^{k-1}\|^2 - \|y-x^{k-1}\| \|A^T B\| (\|x^{k-1}-c\| + \delta)$$

where $\lambda > 0$ is the smallest eigenvalue of $(A^T B + B^T A)/2$. The right hand side of this inequality of positive for sufficiently large $\|y-x^{k-1}\|$. Now suppose $\sigma = [v^0,\ldots,v^{n+1}] \in T^k$ is not contained in $D' \times [0,1]$, where $D' = \{x \in R^n: \|y-x\| \leq \delta$ for some $y \in D\}$. Then if $v^i = (y^i, \varepsilon_i)$, $h(v^i) = f(y^i)$ or $r(y^i) = A(y^i - x^{k-1})$. Since $y^i \notin D$, in either case we find $h(v^i)^T B(x-c) > 0$ for any $x \in [y^0,\ldots,y^{n+1}]$. Thus σ cannot be complete, and the theorem is proved.

Finally we indicate briefly the continuous deformation algorithm of Eaves and Saigal [15]. We define h by

$$h(x,t) = \begin{cases} 2tf(x) + (1-2t)r(x) & 0 \leq t \leq 1/2 \\ f(x) & 1/2 \leq t \leq 1 \end{cases}$$

and use a very special triangulation T of $R^n \times [0,1)$ with the properties:

a) for $k = 0,1,\ldots,$ a subset T_k of T triangulates $R^n \times [1-2^{-k}, 1-2^{-k-1}]$.

b) p mesh $T_k \to 0$.

Notice that this algorithm avoids restarting by using, in effect, a collection of triangulations T_k that "stack up" naturally. Also, the homotopy goes from r to f as t goes from 0 to 1/2 (the first "layer" of the triangulation) and thereafter stays at f. Thus much of the effect of the artificial function r is removed, while in Merrill's algorithm, the matrix A (or a sequence of matrices A_k) must be used at each restart. The algorithm traces a path in $h_T^{-1}(0)$ (or $h_T^{-1}(<\varepsilon>)$). This path necessarily meets an infinite number of simplices, but if condition 6.1 obtains all these lie in a set $D\times[0,1)$ for some bounded D. Thus all limit points lie in $D\times\{1\}$ and the projection on R^n of any of these is a zero of f. We illustrate the algorithm below for $n = 1$, but stress that, for $n > 1$, the sequence of levels penetrated need not be monotonic. Note that, with the triangulation shown, the algorithm for $n = 1$ is merely the bisection algorithm after a zero of f is bracketed.

Example 6.2. Let $n = 1$, $f(x) = x^3+2$ and $r(x) = x - 1/2$. A suitable triangulation T and $h_T^{-1}(0)$ are shown in figure 6.2.

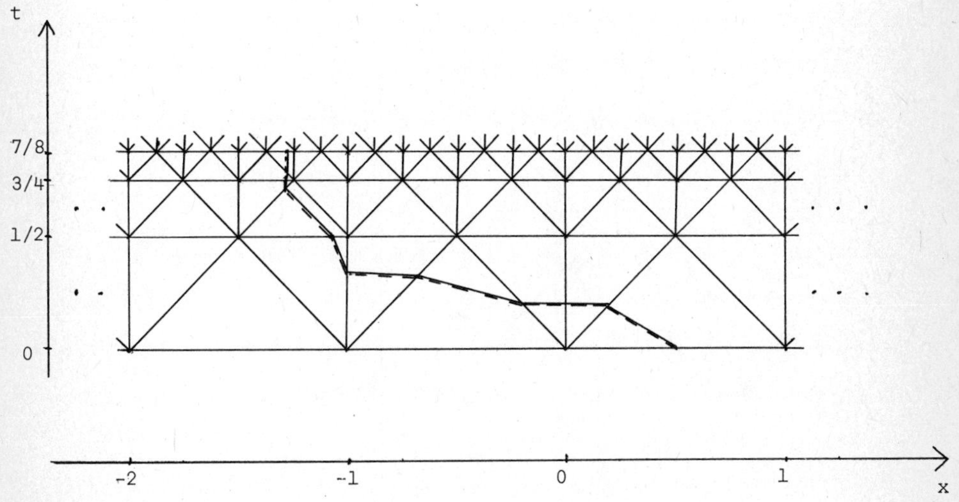

Figure 6.2

7. Triangulations

The algorithms of sections 4 and 6 required the construction of triangulations of arbitrary mesh size (or projected mesh size) and with simple rules for moving between simplices. In this section we describe some.

A. The triangulation J_1 of R^n.

The set J_1^0 of vertices is Z^n, where Z is the set of integers. Each simplex of J_1 depends on an initial vertex v, each of whose components is an odd integer; a permutation π of $\{1,2,\ldots,n\}$; and a "sign" vector $s \in R^n$, each of whose components is ± 1. We denote this simplex $j_1(v,\pi,s)$ and its vertices are $v^0,\ldots,v^n$, where

$$v^0 = v, \quad v^i = v^{i-1} + s_{\pi(i)} e^{\pi(i)}, \quad 1 \leq i \leq n, \tag{7.1}$$

where e^j is as before the jth unit vector. Then J_1 is the set of all such $j_1(v,\pi,s)$.

For $n = 2$, the triangulation J_1 (and its alternate name, "Union Jack triangulation" [79]) is illustrated in figure 7.1. The shaded triangle is $j_1((-1,1)^T, (2,1), (1,-1)^T)$.

For a proof that J_1 is indeed a triangulation of R^n, see e.g. [76].

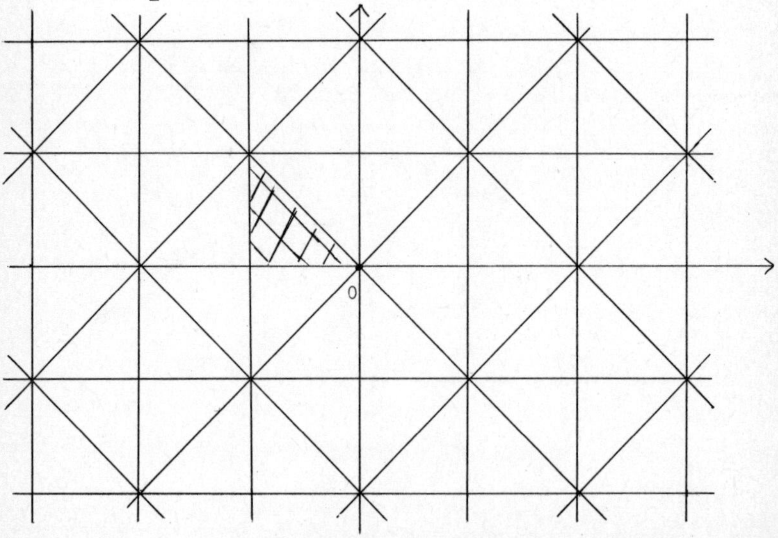

Figure 7.1

In order to move sequentially from simplex to adjacent simplex of J_1 as required by the algorithms, we need the following "pivot rules." Suppose $\sigma = j_1(v,\pi,s) = [v^0,v^1,\ldots,v^n] \in J_1$, where the v^j's are ordered as in (7.1). Let $i \in \{0,1,\ldots,n\}$ be given. The simplex $\sigma' = j_1(v',\pi',s') \neq \sigma$ that contains the facet $[v^0,\ldots,v^{i-1},v^{i+1},\ldots,v^n]$ of σ is then given by the rules of table 7.1.

	v'	π'	s'
$i = 0$	$v-2s_{\pi(1)}e^{\pi(1)}$	π	$s-2s_{\pi(1)}e^{\pi(1)}$
$0 < i < n$	v	$(\pi(1),\ldots,\pi(i+1),\pi(i),\ldots,\pi(n))$	s
$i = n$	v	π	$s-2s_{\pi(n)}e^{\pi(n)}$

Table 7.1

For example, the result of removing v^i from the shaded triangle in figure 7.1, for $i = 0,1$, and 2, yields the triangles $j_1((-1,-1)^T, (2,1), (1,1)^T)$, $j_1((-1,1)^T, (1,2), (1,-1)^T)$ and $j_1((-1,1)^T, (2,1), (-1,-1)^T)$ respectively.

To obtain triangulations of R^n of arbitrarily small mesh size, we may use $\delta J_1 = \{\delta j_1(v,\pi,s)\}$, where for any $S \subseteq R^n$, $\delta S = \{\delta x: x \in S\}$. Note that δJ_1 has mesh size $\sqrt{n}\delta$.

B. The triangulations K_1 and K_1' of R^n

Let us write $k_1(v,\pi)$ for $j_1(v,\pi,e)$ defined as in (7.1), where v can be any point in Z^n. The triangulation K_1 is the set of all such $k_1(v,\pi)$'s. Next let $C^n(s) = \{x \in R^n: s_i x_i \geq 0$ for all $i\}$ be the orthant corresponding to the sign vector s. Then K_1' is the set of all $j_1(v,\pi,s)$'s with v an arbitrary point of $Z^n \cap C^n(s)$. We illustrate K_1 and K_1' in figure 7.2, and merely remark that their pivot rules are about as simple as those of J_1. For details, see [80].

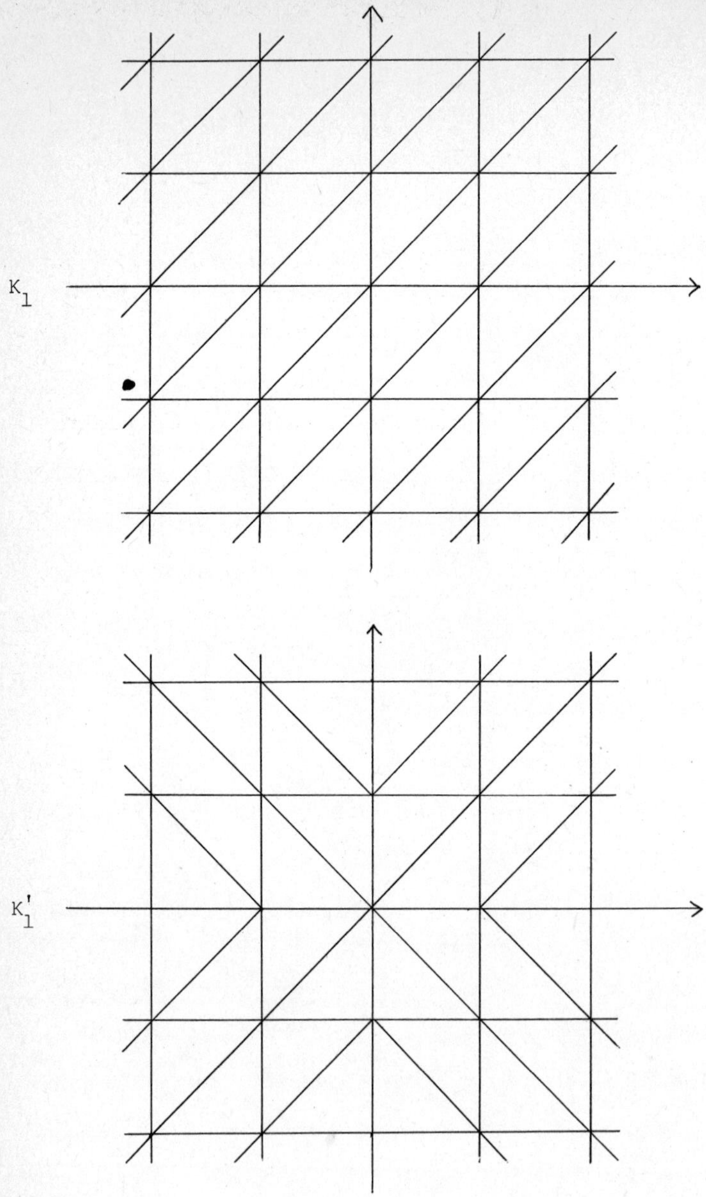

Figure 7.2

C. The triangulation $\tilde{J}_1(\delta)$ of $R^n\times[0,1]$.

To obtain triangulations of $R^n\times[0,1]$ of arbitrary projected mesh size, we remark that if $A: C \to D$ is an affine isomorphism and T a triangulation of C, then $AT = \{A\sigma: \sigma \in T\}$ is a triangulation of D. Letting A_δ be the linear isomorphism of R^{n+1} into itself represented by the matrix $\begin{pmatrix} \delta I & 0 \\ 0 & 1 \end{pmatrix}$, we obtain from the (n+1)-dimensional triangulation J_1 a triangulation $A_\delta J_1$ that takes steps of length $\pm\delta$ in the first n coordinate directions, and ± 1 in the last. Now we use the following.

Theorem 7.1. If $C \subseteq R^m$ is an m-dimensional polyhedron, and each hyperplane H meeting C in an (m-1)-face of C is a union of facets of a triangulation T of R^m, then the simplices of T lying in C form a triangulation $T(C)$ of C.

The proof is straightforward and omitted. The theorem becomes very useful since $x \in R^m$ lies in a facet of J_1 iff x_i is integer for some i or $x_i + x_j$ or $x_i - x_j$ is an even integer for some $i \neq j$. Similar characterizations exist for K_1 and K_1'.

An immediate corollary of theorem 7.1 is that a subset of $A_\delta J_1$ triangulates the polyhedron $R^n\times[0,1]$, whose n-faces are $R^n\times\{0\}$ and $R^n\times\{1\}$. We call this subset $\tilde{J}_1(\delta)$. Note that the projected mesh size of $\tilde{J}_1(\delta)$ is $\sqrt{n}\delta$ and that the induced triangulations of $R^n\times\{0\}$ and $R^n\times\{1\}$ are copies of δJ_1. A similar procedure gives $\tilde{K}_1(\delta)$ and $\tilde{K}_1'(\delta)$. The pivot rules of all are basically as simple as table 7.1; thus in Merrill's restart algorithm, each "step 2" in a major cycle is trivial to perform.

D. The triangulation $J_2(m)$ of S^n.

We briefly mention that the techniques above can also easily give triangulations of $S^n = \{x = (x_0,x_1,\ldots,x_n)^T: x \geq 0, \sum_j x_j = 1\}$ of arbitrary mesh size. For instance, a subset of J_1 triangulates the simplex $\{x \in R^n: m \geq x_1 \geq x_2 \geq \ldots \geq x_n \geq 0\}$ for any positive integer m. Then the affine isomorphism A defined by $Ax = m^{-1}(m-x_1,x_1-x_2,\ldots,x_{n-1}-x_n,x_n)^T$ takes this simplex onto S^n and therefore yields a triangulation $J_2(m)$ of S^n. The mesh size of this triangulation is

$\sqrt{4n-2}/m$. Similarly one can construct $K_2(m)$, of mesh about $\sqrt{n}/m$. (Figure 4.1 shows $K_2(5)$.)

However, in [80] it is suggested that an orthogonal transformation A instead of that above yields more efficient triangulations of the affine hull of S^n, at least for n bigger than about 10. Computational experience cited in [80] tends to support this view, especially for the J-based triangulations.

E. The triangulation J_3 of $R^n \times (0,1]$.

This is a triangulation that, after substituting $1-t$ for t, satisfies the requirements for the continuous deformation algorithm of Eaves and Saigal [15]. We will not describe it in detail here--see e.g. [76]. We merely note that a subset of the facets of J_3 gives a triangulation of $R^n \times \{2^{-k}\}$ that is a copy of $2^{-k}J_1$ so that the induced mesh size decreases by a factor of 2 in each layer. Also, the pivot rules, while having many more cases than those of J_1, require very little more work to perform. Figure 6.2 shows (an inverted version of) J_3 for $n = 1$. It seems that J_3 is far more efficient than K-based continuous deformation triangulations [77,80].

F. Recent Work

Instead of using affine isomorphisms as above, we might ask which triangulation AJ_1, AK_1, or AK_1' is most "efficient", where A varies over affine isomorphisms of R^n onto itself. For J_1 and K_1', it appears that the original versions are near optimal, whereas improvements to K_1 can be made. Building on work of Todd [80], van der Laan and Talman [48] have constructed a particularly good transformation of K_1 and verified its advantages computationally. In the next section we describe some other important contributions of these authors.

Several attempts have been made to decrease the mesh size for a continuous deformation triangulation by a factor of more than two in each layer; see van der Laan and Talman [49], Shamir [72], Barany [5], Kojima and Yamamoto [40] and Van der Heyden [94]. A cautionary note is added by Todd [82] and Todd and Acar [91], who show that attempts to decrease the mesh size too fast may force a large number of iterations.

8. An Artificial Cube Algorithm

Each major cycle of the restart algorithm traverses a sequence of (n+1)-dimensional simplices. As each new simplex is entered, we must evaluate h at the new vertex, i.e., evaluate f or r, and then perform a linear programming pivot step. If v lies in $R^n \times \{1\}$ we evaluate f, thus gaining new information about the problem of interest; but if v lies in $R^n \times \{0\}$ we must still perform the linear algebra step, which seems rather wasted since we have gained no new information. In this section we discuss an approach of van der Laan and Talman designed to reduce the number of such undesirable steps.

The basic idea is that the hyperplane $R^n \times \{0\}$ is only introduced to provide a start for the algorithm--so such a method might be much more efficient if $R^n \times \{0\}$ were replaced by a much smaller set that is subdivided with very few vertices. In [45,46], van der Laan and Talman replaced $R^n \times \{0\}$ with an artificial simplex, while in [51] they discussed a whole family of algorithms ranging from this artificial simplex algorithm to another in which an artificial cube replaced $R^n \times \{0\}$. We will discuss only this latter method. A very similar algorithm using integer labelling was developed independently by Reiser [60]. Related algorithms were introduced by Tuy, Thoai and Muu [93] and Van der Heyden [94]. The original descriptions avoided the introduction of the extra dimension altogether, but this disguises the actual homotopy used. Once the geometric interpretation was provided in Todd [81] and Todd and Wright [92] (see also Todd [83], van der Laan and Talman [47]) other possibilities became apparent, for instance the octahedral algorithm of Wright [98]. A unifying framework and analysis is provided in Kojima and Yamamoto [39,40], Kojima [37], and Freund [21].

For $C,D \subseteq R^n$, let us write $C \# D$ for $\mathrm{conv}((C\times\{0\}) \cup (D\times\{1\}))$. Let B^n_∞ be the cube $\{x \in R^n: \|x\|_\infty \leq 1\}$, where $\|x\|_\infty = \max_i |x_i|$ and $\|x\|_1 = \sum_i |x_i|$ as usual. The algorithm uses a "subdivision" of $B^n \# R^n$ to approximate the homotopy h of (5.3) with $x^0 = 0$. (We translate the function f if necessary so that $x^0 = 0$ is an appropriate starting point.) The quotes are necessary since the "subdivision" is not locally finite.

Recall the orthants $C^n(s)$ corresponding to the sign vectors $s \in R^n$, introduced in the previous section. We extend that notion here, calling a sign vector in R^n any $(0,\pm 1)$-vector s, and defining

$$C^n(s) = \{x \in R^n: \|x\|_1 = s^T x\}.$$

Thus $C^n(s)$ is an orthant of some coordinate subspace of R^n. Theorems 3.1 and 7.1 imply that subsets of J_1^+, K_1^+ and $K_1'^+$ triangulate each $C^n(s)$. Let T be a triangulation of R^n that also triangulates in this way all orthants $C^n(s)$. If s has k nonzero components, $C^n(s)$ has dimension k, and a simplex $\sigma \in T^k$ that lies in $C^n(s)$ will be called a full-dimensional simplex in $C^n(s)$. In some sense dual to the cone $C^n(s)$ is the face

$$B_\infty^n(s) = \{x \in B_\infty^n: s^T x = s^T s\}$$

of B_∞^n; it has dimension $n-k$ and its tangent space (i.e. the subspace spanned by differences of its elements) is the orthogonal complement of the tangent space of $C^n(s)$. Then let L be the collection of all polyhedra (or cells) $\rho \# \sigma$, where $\rho = B_\infty^n(s)$ and σ is a full-dimensional simplex of $C^n(s)$ for some sign vector s. Some of these cells are illustrated for $n = 2$ in figure 8.1. It is not hard to establish

Proposition 8.1.

a) Each cell of L is $(n+1)$-dimensional.

b) Let τ be a facet (n-face) of a cell of L. Then either i) τ lies in $R^n\times\{1\}$ or is $B_\infty^n\times\{0\}$, and is a facet of precisely one cell of L, or ii) τ is a facet of precisely two cells of L.

c) If $C \subseteq R^n$ is compact, there is a finite number of cells of L contained in $B_\infty^n \# C$.

d) If h is defined by (5.3) then there is exactly one PL approximation h_L to h with respect to L, i.e., h_L agrees with h on each vertex of each cell of L and is linear in each such cell.

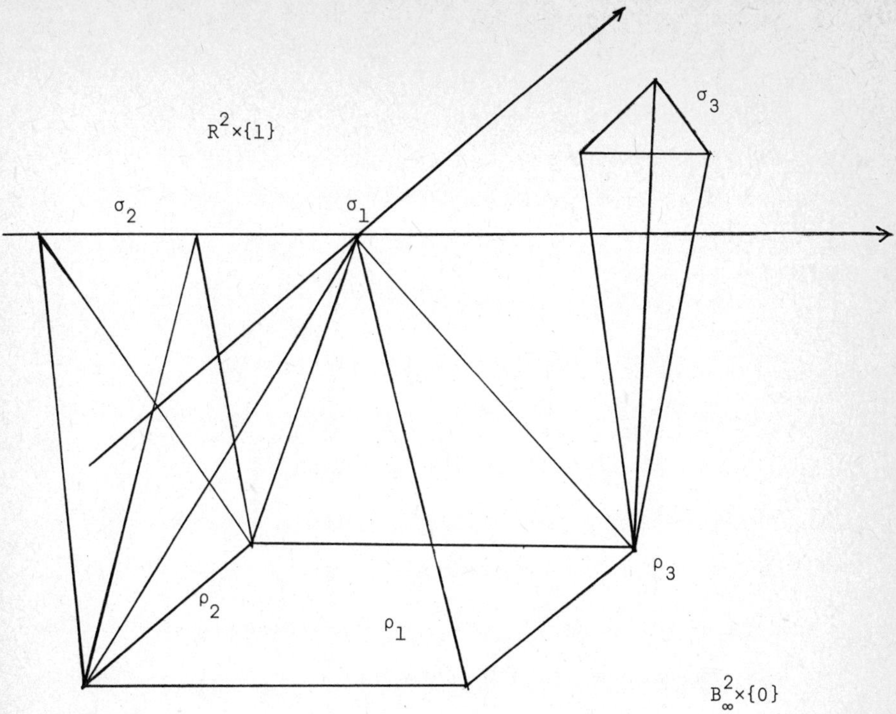

Figure 8.1

For part (d), uniqueness is easy by considering the difference of two PL approximations. Existence is harder since the cells are not necessarily simplices and so the representation of a point as a convex combination of vertices of the cell $\rho \# \sigma$ is not unique. However, the orthogonality of the tangent spaces of ρ and σ implies that any point in $\rho \# \sigma$ is a unique convex combination of a point in $\rho\times\{0\}$ and a point in $\sigma\times\{1\}$. Since r is linear we can define h_L thus from r and f_T and confirm that it is linear in each cell. Note that (due to lack of local finiteness) h_L is generally not continuous at points in $\partial B^n_\infty\times\{0\}$.

Our aim is to trace the path in $h_L^{-1}(0)$ from (0,0). To do this we need an analogue of theorem 6.1.

Theorem 8.1 [83]. Let $\rho = B^n_\infty(s)$ and $\sigma = [y^0,\dots,y^j] \in T$, $\sigma \subseteq C^n(s)$ for some sign vector s with j nonzero elements. Let A_s denote the submatrix of A corresponding to indices k with $s_k = 0$; A_s thus has $n-j$ columns. Then $h_L^{-1}(0)$ meets $\rho \# \sigma$ iff there is a feasible solution to

$$\begin{bmatrix} 1 & \dots & 1 & 0 & 1 \\ f(y^0) & & f(y^j) & A_s & As \end{bmatrix} w = \begin{pmatrix} 1 \\ 0 \end{pmatrix}, \tag{8.1}$$

$$w_i \geq 0, \quad 0 \leq i \leq j;$$

$$w_{n+1} \geq w_k \geq -w_{n+1}, \quad j < k \leq n;$$

$$w_{n+1} \geq 0 \quad \text{if} \quad j = n.$$

Note that the linear system (8.1) has $n+2$ unknowns and $n+1$ equations, with up to $2n+1$ inequalities. The large number of inequalities is not surprising, since some of the cells of L have $2n+1$ facets, and there must be an inequality corresponding to each facet to indicate whether the path in $h_L^{-1}(0)$ goes through that facet. The description of this algorithm by van der Laan and Talman [51] used a linear system of size $2n\times(2n+1)$, while Kojima and Yamamoto [40] show that a system of size $n\times(n+1)$ suffices--they also have simple rather than "variable" upper bounds on the variables.

It should be clear how a major cycle of the algorithm is performed. We translate the function if necessary so that $x^0 = 0$ is a suitable starting point and then define h as in (5.3). We choose a triangulation T, e.g., δJ_1. Now there is clearly a feasible solution to (8.1) with $s = 0$; $w = (0,0,\dots,0,1)^T$ works. Note that $\sigma = [y^0]$ with $y^0 = 0$. Next $f(y^0)$ is evaluated and the weight w_0 on the corresponding column in (8.1) is increased. Thus another solution to (8.1) with a different inequality, say $w_k \geq -w_{n+1}$, tight. Now if the corresponding column of A_s is subtracted from $As = 0$, the result is At where $t = -e^k$. We proceed by finding the simplex $\sigma' = [y^0,y^1] \subseteq C^n(t)$ of T that has $\sigma = [y^0]$ as a facet. The theorem then shows that $\rho' \# \sigma'$, with

$\rho' = B_\infty^n(t)$, contains a zero of h_L. We continue in this way, alternating pivots in the linear system (8.1) with "pivots" in the subdivision L. The major cycle ends when the weight w_{n+1} goes to zero.

We omit further details and conclude with the following remarks. First, global convergence of the artificial simplex and artificial cube algorithms is guaranteed if f satisfies condition 6.1, see [81,83] and van der Laan and Talman [50]. However, the conditions of the Leray-Schauder theorem which, after mild strengthening, suffice for the convergence of the algorithms of section 6 (see, e.g., Saigal [63]) are not sufficient for these algorithms. Secondly, the pivots in the linear system (8.1) can be carried out in an analogous way to those in standard nonnegativity-constrained systems. However, various special techniques are useful if factorization of basis matrices is used, as is advisable for numerical stability. In particular, the column of an upper triangular factor corresponding to the final column in (8.1) should always be the last column, and new columns should therefore be introduced in the penultimate position. See [86].

9. Acceleration

One of the attractive features of the class of piecewise-linear homotopy algorithms is that they can be implemented to converge quadratically to a solution of smooth regular problems. This possibility was first raised by Saigal [64]; see also [67,83]. While it is possible to implement continuous deformation algorithms to converge fast locally [67], it is much simpler with a restart algorithm. We therefore consider Merrill's algorithm to compute a zero of $f: R^n \to R^n$, but use a method of analysis [83] that can be applied to several other algorithms, e.g. the artificial simplex [81], cube [83] and octahedron [98] methods.

Suppose a sequence of triangulations $\{T_k\}$ is used, with projected mesh sizes $\{\delta_k\}$ tending to 0. Let $\delta = \max_k \delta_k$, and assume that all zeroes of PL approximations to f with respect to triangulations of mesh size at most δ lie in a compact set $\hat{C} \subseteq R^n$ (e.g., condition 6.1 suffices).

Let $C = \{x \in R^n: \|x-y\| \leq \delta$ for some $y \in \hat{C}\}$ and assume f is continuously differentiable on C and its derivative Df satisfies a Lipschitz condition on C:

$$\|Df(x) - Df(y)\| \leq \kappa \|x-y\| \quad \text{for all} \quad x,y \in C. \tag{9.1}$$

This is our smoothness assumption. Next we assume that $Df(x^*)$ is nonsingular at each zero x^* of f. Together with the fact that all zeroes of f lie in $\hat{C}$, this implies that there is a finite number of them, so we can alternatively suppose there exists $\mu \geq 1$ with

$$\|(Df(x^*))^{-1}\| \leq \mu \quad \text{for all zeroes} \quad x^* \text{ of } f. \tag{9.2}$$

This is our regularity assumption.

The key to acceleration is the realization that each major cycle produces not only an approximate zero x^k of f but also an approximation to $Df(x^k)$, which we denote D_k. Indeed we have evaluated $f(y^{ki})$, $i = 0,\ldots,n$, where $[y^{k0},\ldots,y^{kn}]$ is the simplex of $T_k(1)$ containing x^k. Then D_k is uniquely defined by $D_k(y^{ki}-x^k) = f(y^{ki})$, $i = 0,1,\ldots,n$. In particular, suppose that $T_k = \mathcal{J}_1(\hat{\delta}_k)$, $\hat{\delta}_k = \delta_k/\sqrt{n}$. Then each $y^{ki}-y^{ki-1}$ is a multiple of a unit vector, so that D_k can be trivially obtained column by column. We again require the useful inequality, e.g., Ortega and Rheinboldt [58] p. 73: Given (9.1),

$$\|f(w) - f(z) - Df(z)(w-z)\| \leq \frac{1}{2}\kappa \|w-z\|^2 \tag{9.3}$$

for all w,z in C. Using equation (9.3) we easily obtain the following:

<u>Proposition 9.1.</u> $\|f(x^k)\| \leq \frac{1}{2}\kappa\delta_k^2$ and $\|D_k - Df(x^k)\| \leq \kappa n\delta_k$.

The proposition implies that for large enough k, D_k is nonsingular. Indeed, we use the result (cf. [58], p. 45):

<u>Lemma 9.1.</u> Let B and E be real $n \times n$ matrices. Suppose B is nonsingular with $\|B^{-1}\| \leq \beta$ and $\|E\| \leq \hat{\gamma}$. Then if $2\beta\gamma \leq 1$, $B+E$ is nonsingular with $\|(B+E)^{-1}-B^{-1}\| \leq 2\beta^2\gamma$.

(Note that the coefficients "2" in the above were missing in [83], causing several minor errors.)

From this it is simple to establish (using contradiction for the first part)

Proposition 9.2. For any $\varepsilon > 0$ we can choose k_0 such that each x^k for $k \geq k_\sigma$ lies within ε of some zero of f. Further k_0 can be chosen so that $\|D_k^{-1}\| \leq 2\mu$ and $\|D_k^{-1}f(x^k)\| \leq \mu\kappa\delta_k^2 \leq \delta_k/2$ for all $k \geq k_0$.

Now we can state the modification of the algorithm:

Let $\bar{x}^0 = x^0$, $A_0 = A$ and $T_1 = \mathcal{J}_1(\hat{\delta}_1)$, $\delta_1 = \sqrt{n}\,\hat{\delta}_1$.

Iteration k: Let $x^k = M(f, \bar{x}^{k-1}, A_{k-1}, T_k)$ and $D_k \approx Df(x^k)$ be obtained as above. Stop if $f(x^k) = 0$.

If D_k is singular or

$\|s^k\| > \delta_k/2$, where $s^k = -D_k^{-1}f(x^k)$, or

$\|f(x^k+s^k)\| > \|f(x^k)\|/2$

set $\bar{x}^k = x^k$, $A_k = A$, $\delta_{k+1} = \delta_k/2$, $\hat{\delta}_{k+1} = \hat{\delta}_k/2$ and $T_{k+1} = \mathcal{J}_1(\hat{\delta}_{k+1})$.

Otherwise, set $\bar{x}^k = x^k+s^k$, $A_k = D_k$, $\delta_{k+1} = \|s^k\|$, $\hat{\delta}_{k+1} = \delta_{k+1}/\sqrt{n}$ and T_{k+1} the translation of $\mathcal{J}_1(\hat{\delta}_{k+1})$ by the vector $(\bar{x}^k - \hat{\delta}_{k+1}(n+1)^{-1} \cdot (n,n-1,\ldots,1)^T, 0)$. (The effect is to put $(\bar{x}^k, 0)$ into the center of a facet of T_{k+1}.)

We will assume that each x^k exists, i.e., each major cycle converges. Note that, while f may satisfy condition 6.1 with matrix A, this condition may fail with A replaced by D_k. There are methods to maintain global convergence while accelerating, but they are rather complicated (see, e.g., [67]).

Our aim is to sketch a proof of

Theorem 9.1. The sequence x^k converges Q-quadratically to some zero x^* of f, i.e., for some $\rho > 0$ and all k,

$$\|x^{k+1} - x^*\| \leq \rho\,\|x^k - x^*\|^2. \tag{9.4}$$

Moreover, for sufficiently large k, each x^{k+1} is obtained from $\bar{x}^k$ in exactly $n+1$ evaluations of f and linear programming pivot steps.

Note first that proposition 9.2 implies that the first two conditions for acceleration are satisfied for all sufficiently large k. The third condition follows easily from (9.3) for all sufficiently large k. Thus the algorithm eventually accelerates at every iteration.

When we use $A_k = D_k$, f is very close to the artificial function

$$r^k(x) = D_k(x-\bar{x}^k). \tag{9.5}$$

Our strategy is to analyze the sequence of simplices met when $f = r^k$, and then show that f is sufficiently close to r^k that exactly the same sequence is in fact encountered.

Let $y^0 = \bar{x}^k + \hat{\delta}_{k+1}(n+1)^{-1}(1,2,\ldots,n)^T$ and $[y^0,y^1,\ldots,y^n] = j_1(y^0,(n,n-1,\ldots,1),-e$ Let $v^i = (y^i,0)$ and $\bar{v}^i = (y^i,1)$ for $i = 0,1,\ldots,n$. Then $\tau_0 = [v^0,v^1,\ldots,v^n]$ is the facet of T_{k+1} containing $(\bar{x}^k,0)$. Define $\tau_i = [\bar{v}^0,\bar{v}^1,\ldots,\bar{v}^i,v^{i+1},\ldots,v^n]$ for $i = 1,2,\ldots,n$. Notice that if $f = r^k$, the label matrix of each τ_i is the same, and is

$$L_{\tau_0} = B \equiv \begin{pmatrix} 1 & 0 \\ 0 & \hat{\delta}_{k+1}I \end{pmatrix}\begin{pmatrix} 1 & 0 \\ 0 & D_k \end{pmatrix}\begin{pmatrix} 1 & \cdots & 1 \\ \frac{y^0-\bar{x}^k}{\hat{\delta}_{k+1}} & \cdots & \frac{y^n-\bar{x}^k}{\hat{\delta}_{k+1}} \end{pmatrix} \tag{9.6}$$

$$\equiv \begin{pmatrix} 1 & 0 \\ 0 & \hat{\delta}_{k+1}I \end{pmatrix}\hat{B},$$

say, whose inverse is

$$B^{-1} = \begin{bmatrix} \frac{1}{n+1} & \bigcirc & & 1 \\ \vdots & & \cdot^{\cdot^{\cdot}} & -1 \\ \frac{1}{n+1} & 1 & \cdot^{\cdot^{\cdot}} & \\ \frac{1}{n+1} & -1 & & \bigcirc \end{bmatrix}\begin{pmatrix} 1 & 0 \\ 0 & D_k^{-1} \end{pmatrix}\begin{pmatrix} 1 & 0 \\ 0 & \hat{\delta}_{k+1}^{-1}I \end{pmatrix} \tag{9.7}$$

$$= \hat{B}^{-1}\begin{pmatrix} 1 & 0 \\ 0 & \hat{\delta}_{k+1}^{-1}I \end{pmatrix}.$$

Note that $\hat{B}^{-1}$ is lexicographically nonnegative (its first column is $(n+1)^{-1}e$) and its norm is bounded by $2(n+1)\,\|D_k^{-1}\| \le 4(n+1)\mu$ for $k \ge k_0$. Since the label matrix of each τ_i has a lexicographically nonnegative inverse, we have

Lemma 9.2. If $f = r^k$, the kth major cycle generates the sequence $\tau_0,\tau_1,\ldots,\tau_n$ of very complete n-simplices and produces x^{k+1} in exactly n+1 evaluations of f (at $y^0,y^1,\ldots,y^n$) and exactly n+1 linear programming pivot steps.

Next, the bounds we have obtained in proposition 9.1 and the ubiquitous inequality (9.3) easily yield

Lemma 9.3. For any $\theta > 0$, there is an index $k_1 = k_1(\theta) \ge k_0$ such that for $k \ge k_1$ and $\|x-\bar{x}^k\| \le \sqrt{n}\,\hat{\delta}_{k+1}$,

$$\|f(x) - D_k(x-\bar{x}^k)\| \le \theta\hat{\delta}_{k+1}.$$

Now we can prove

Proposition 9.3. For sufficiently large k, the algorithm obtains x^{k+1} from $\bar{x}^k$ in exactly n+1 evaluations of f and linear programming pivot steps.

Proof. We show that the n-simplices $\tau_0,\tau_1,\ldots,\tau_n$ remain very complete even if $f \neq r^k$. Let B_i denote the label matrix of τ_i, so that

$$B_i = \begin{bmatrix} 1 & \cdots & 1 & 1 & \cdots & 1 \\ f(y^0) & & f(y^i) & r(y^{i+1}) & & r(y^n) \end{bmatrix}$$

$$= \begin{pmatrix} 1 & 0 \\ 0 & \hat{\delta}_{k+1}I \end{pmatrix} \hat{B}_i,$$

say. Lemma 9.3 states that each column of $\hat{B}_i$ is within θ of the corresponding column of $\hat{B}$ for $k \ge k_1(\theta)$. We will choose $\theta = 1/64(n+1)^4\mu^2$. Then $\|\hat{B}_i-\hat{B}\| \le 1/64(n+1)^3\mu^2$, so that by lemma 9.1, $\hat{B}_i$ is nonsingular and $\|\hat{B}_i^{-1}-\hat{B}^{-1}\| \le 2\cdot(4(n+1)\mu)^2/64(n+1)^3\mu^2 = 1/2(n+1)$. It follows that each element of $\hat{B}_i^{-1}$ is within

$1/2(n+1)$ of the corresponding element of $\hat{B}^{-1}$, and in particular that the first column of $\hat{B}_i^{-1}$ is positive. This shows that τ_i is very complete, and hence that the algorithm generates precisely this sequence of facets.

Proposition 9.3 implies that $\{x^k\}$ converges to a single zero x^* of f, since for large k, $\|\bar{x}^k - x^k\| \leq \delta_{k+1}$ and $\|x^{k+1} - \bar{x}^k\| \leq \delta_{k+1}$; since $\{\delta_k\} \to 0$ at least linearly, we have a Cauchy sequence. Next we have

$$\|f(x^{k+1})\| \leq \frac{1}{2}\kappa\delta_{k+1}^2 \leq \frac{1}{2}\kappa\,\|D_k^{-1}\|^2\,\|f(x^k)\|^2$$

$$\leq 2\kappa\mu^2\,\|f(x^k)\|^2$$

for sufficiently large k. Since $f(x) \approx Df(x^*)(x-x^*)$ for x close to x^*, there exist constants $0 < \alpha < \beta$ with

$$\alpha\,\|f(x)\| \leq \|x-x^*\| \leq \beta\,\|f(x)\|$$

for x close enough to x^*. Thus since the function values converge quadratically to 0, $\{x^k\}$ converges quadratically to x^*. This proves theorem 9.1.

10. Exploiting Special Structure

We stated in the introduction that piecewise-linear homotopy algorithms tend to be slower than more classical methods in solving systems of smooth equations, at least when a good starting point is available. One reason for this is the large number of simplices or cells that may be traversed. Indeed, each of the triangulations considered in Section 7 divides each unit cube into $n!$ simplices. Of course, homotopy algorithms usually traverse a very small fraction of these (we will discuss this further in the next section), but each major cycle may still contain a fair number of iterations, each involved with the traversal of a simplex. This is particularly true when a point-to-set mapping F is involved.

The first point we wish to make in this section is that frequently the "pieces of linearity" of h_T are cells much larger than individual simplices of T. Indeed, if h is given by (5.3), the linearity of r automatically generates larger cells. Consider again example 6.1. Here h_T is in fact linear in the union of the first three triangles met by the algorithm. It can easily be seen that h_T will be linear in the first two of these for any function f, solely because of the linearity of r. The result is that Merrill's algorithm can be implemented with larger pieces of linearity. Indeed, if the triangulation $T = \tilde{J}_1(1)$ of $R^n \times [0,1]$ is used, there are $2^n(n+1)!$ simplices in $[0,2]^n \times [0,1]$, but fewer than $e^{1/2}2^n n!$ pieces of linearity of h_T [85]. Note that $2^n n!$ of these correspond to the facets of $T(1)$ triangulating $[0,2]^n \times \{1\}$ and thus providing a PL approximation of f. Thus, in such an implementation, the extra homotopy dimension incurs a minimal penalty on the efficiency of the algorithm. Indeed, the pieces of linearity bear a striking resemblance to those that arise in the subdivision L corresponding to the "artificial cube" algorithm--see [85].

Kojima [36] first exploited the special structure of functions to devise more efficient homotopy algorithms; see also Kojima [35]. The description above is based on [85], where the idea of large pieces was systematically developed. As we have seen, large pieces result from just the linearity of r. Even larger pieces result when f has some additional structure, e.g. (partial) separability (Kojima [36], Todd [84,85]), partial linearity [85] or sparsity [84,89]. We illustrate this by considering separability.

Let $f: R^n \to R^n$ be separable, i.e. $f(x) = \sum_i f^i(x_i)$ where $f^i: R \to R^n$ is continuous for $i = 1,2,\ldots,n$. Let h be defined by (5.3) and let $T = \tilde{J}_1(1)$ (we choose $\delta = 1$ merely for simplicity). Then the pieces of linearity of h_T can be characterized as follows. Let $z \in Z^n$ and let I_e (I_o) be the index set of even (odd) coordinates of z. Let s be a (± 1) sign vector. Define $\rho = \rho(z,s) = \{x \in R^n: x_i = z_i$ for $i \in I_e$, $z_i - 1 \le x_i \le z_i + 1$ for $i \in I_o\}$ and $\sigma = \sigma(z,s) = \{x \in R^n: 0 \le s_i(x_i - z_i) \le 1$ for $i \in I_e$, $x_i = z_i$ for $i \in I_o\}$. Then each piece of linearity is of the form $\rho \# \sigma$ for some z,s [85]. Next, let $B = (b^1,\ldots,b^n)$ be an $n \times n$ matrix with $b^i = a^i$ (the ith column of A in (5.3)) if $i \in I_0$ and

$$b^i = f(z+s_i e^i) - f(z) = f^i(z_i+s_i) - f^i(z_i)$$

if $i \in I_e$. Then we have

Theorem 10.1 [85]. The piece of linearity $\rho \# \sigma$ above contains a zero of h_T iff there is a feasible solution to

$$\begin{bmatrix} 1 & 0 & 1 \\ f(z) & B & r(z) \end{bmatrix} w = \begin{pmatrix} 1 \\ 0 \end{pmatrix}$$

$$w_0 \geq 0; \tag{10.1}$$

$$w_{n+1} \geq 0;$$

$$w_0 \geq w_i \geq 0 \quad \text{for} \quad i \in I_e;$$

$$w_{n+1} \geq w_i \geq -w_{n+1} \quad \text{for} \quad i \in I_0.$$

For details on how to use this theorem in an efficient algorithm, see [85]. Kojima's approach [36] was to write $r(x) = \sum r^i(x_i)$ and then make PL approximations to each $h^i(x_i,t) = tf^i(x_i) + (1-t)r^i(x_i)$; surprisingly, one of his methods turns out to be identical to that above [84]. However, Kojima's linear systems have at least $2n$ equations, while (10.1) has only $n+1$. Note also that if f is sparse, e.g. its derivative matrix $Df(x)$ has (known) sparsity, and if r is chosen to have the same sparsity pattern, B is sparse and thus well-known sparsity techniques can be used in (10.1) (see [86]). Even if f is not separable, sparsity can be exploited [84,89]. As a final remark, this approach to exploiting structure requires the pieces of linearity to be determined and possibly new codes to be developed for each structure. Other methods, in particular smooth continuation methods, can exploit structure much more easily and naturally.

Next we discuss briefly the recent work of Awoniyi and Todd [3] in exploiting structure in problems involving point-to-set mappings. It should be clear that homotopy algorithms generally perform much more poorly when applied to such problems than when searching for zeroes of continuous functions. However, most instances of point-to-set problems lead to mappings F constructed as in section 2.B(ii) from smooth functions. That is, there are regions X_i, $i = 0,1,2,\dots,m$, constructed from smooth functions $g_i : R^n \to R$ by $X_i = \{x \in R^n: g_i(x) = \max_{0\le j\le m} g_j(x)\}$ and smooth functions $f^i: X_i \to R^n$, with

$$F(x) = \text{conv}\{f^i(x): i \in \{0,1,\dots,m\} \text{ with } x \in X_i\}.$$

For such mappings, acceleration techniques can be devised.

Suppose for simplicity that each g_i is affine. Then the X_i's are polyhedral If a restart algorithm is applied to F, we will obtain $\{x^k\}$ with $x^k \in [y^{k0},\dots,y^{kn}]$, a simplex of $T_k(1)$. We can then calculate $I_k = \{i \in \{0,1,\dots,m\}: y^{kj} \in X_i$ for some $j\}$. We might guess that a solution lies in $X_{I_k} = \cap\{X_i: i \in I_k\}$. In this case we project x^k to $\bar{x}^k$ in aff X_{I_k}, use an orthogonal transformation of the triangulation so that aff X_{I_k} resembles a coordinate subspace, and choose $H_T(y,1)$, for $y \in X_{I_k}$, as a particular member of $F(y)$ that lies if possible in the tangent space of X_{I_k}. The problem then resembles that of finding a zero of a smooth function defined in a lower dimensional subspace. The technique is illustrated in figure 10.1, where an arrow points from a point x to $x + H_T(x,1)$. Details may be found in Awoniyi and Todd [3]. Under suitable conditions, superlinear convergence can be achieved (quadratic if each g_i is affine).

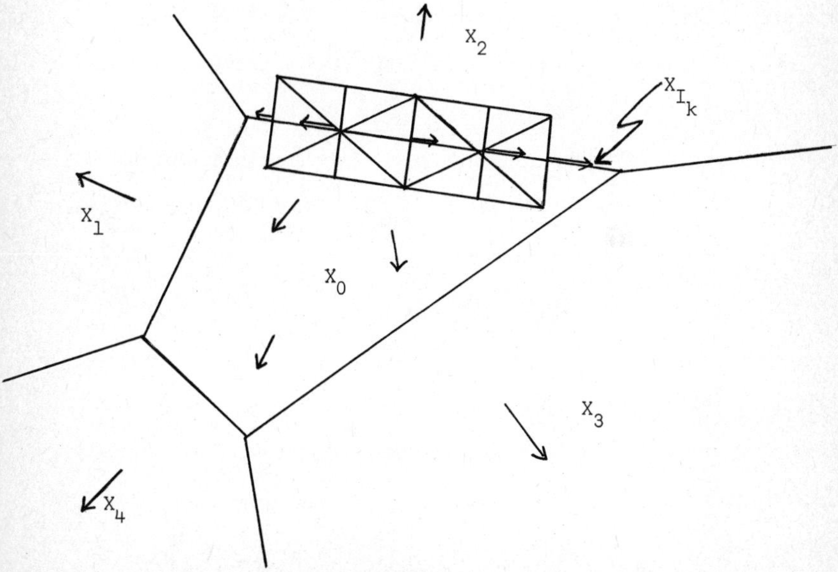

Figure 10.1

11. Remarks on Computational Experience

Here we will merely indicate some papers containing computational results using (piecewise-linear) homotopy algorithms and state some general conclusions. There are two reasonably well-documented piecewise-linear codes available (Saigal [66] and Todd [90]), and Watson and Fenner [97] have published a code using a smooth continuation method.

Computational experience with economic equilibrium problems is reported in Scarf with Hansen [71], van der Laan and Talman [45], Shamir [72] and Todd [80,88]. For constrained optimization problems, see Merrill [55], Fisher and Gould [17], Gochet, Loute and Solow [28], Netravali and Saigal [57], Saigal [64,65], Solow [75] and Awoniyi and Todd [3]. There are also results for other test problems in Saigal [64], van der Laan and Talman [46,48,51] and Allgower and Georg [2]. For smooth continuation methods, see Kellogg, Li and Yorke [34] and Watson [96].

The conclusion from these reports is that the computational effort to solve a problem of dimension n using piecewise-linear homotopy algorithms is of order n^2 function evaluations and linear programming pivot steps (hence order n^4 floating point operations). (However, the author [87] has demonstrated that most algorithms can require an exponential number of steps to solve a system of linear equations if the artificial function is not carefully chosen.) There is a great deal of variability caused by the smoothness or lack of it in the problem, and problems with point-to-set mappings require much more work. For example, the Scarf exchange economy examples of dimensions 4, 7 and 9 require about 50, 100 and 50 function evaluations for all modern algorithms. The 20 dimensional problem of Kellogg, Li and Yorke [34] that is very smooth requires only 80 for PLALGO [90]. However, the 13 dimensional production economy example of Hansen (see [71]) requires 1000-2000 function evaluations without acceleration, and 500-1000 with the devices of section 10 [3]; this example gives rise to a point-to-set mapping.

For general systems $f(x) = 0$ there is some difficulty with the choice of a suitable artificial function r, and some well-known test problems cannot be solved because piecewise-linear approximations have no zeroes (as in the example $f(x) = x^2$ in R^1). Thus it seems that, while acceleration techniques have been very succesful in making the algorithms more efficient, piecewise-linear homotopy methods are most appropriate for highly nonlinear problems of relatively low dimension when suitable boundary conditions can be verified.

References

1. J.C. Alexander and J.A. Yorke, "The homotopy continuation method: Numerically implementable topological procedures," Trans. Amer. Math. Soc. 242, 271-284 (1978).

2. E. Allgower and K. Georg, "Simplicial and continuation methods for approximating fixed points and solutions to systems of equations," SIAM Rev. 22, 28-85 (1980).

3. S.A. Awoniyi and M.J. Todd, "An efficient simplicial algorithm for computing a zero of a convex union of smooth functions," School of Operations Research and Industrial Engineering, Cornell University, Ithaca, New York (1981).

4. I. Barany, "Borsuk's theorem through complementary pivoting," Math. Program. 18, 84-88 (1980).

5. I. Barany, "Subdivisions and triangulations in fixed point algorithms," preprint, International Research Institute for Management Science, Moscow (1979).

6. F.J. Branin, Jr., "Widely convergent method for finding multiple solutions of simultaneous nonlinear equations," IBM J. Res. Develop. 16, 504-522 (1972).

7. A. Charnes, C.B. Garcia and C.E. Lemke, "Constructive proofs of theorems relating to F(x) = y, with applications," Math. Program. 12, 328-343 (1977).

8. S.N. Chow, J. Mallet-Paret and J.A. Yorke, "Finding zeros of maps: Homotopy methods that are constructive with probability one," Math. Comput. 32, 887-899 (1978).

9. D.I.A. Cohen, "On the Sperner lemma," J. Comb. Theory 2, 585-587 (1967).

10. G.B. Dantzig, Linear programming and extensions, Princeton University Press, Princeton, NJ (1963).

11. F.-J. Drexler, "A homotopy method for the calculation of all zeros of zero-dimensional polynomial ideals," in Continuation methods, H. Wacker, ed. 69-94, Academic Press, New York (1978).

12. B.C. Eaves, "Computing Kakutani fixed points," SIAM J. Appl. Math. 21, 2, 236-244 (1971).

13. ______ "Homotopies for computation of fixed points," Math. Program. 3, 1-22 (1972).

14. ______ "A short course in solving equations with PL homotopies," SIAM-AMS Proc. 9, 73-143 (1976).

15. B.C. Eaves and R. Saigal, "Homotopies for computation of fixed points on unbounded regions," Math. Program. 3, 225-237 (1972).

16. B.C. Eaves and H. Scarf, "The solution of systems of piecewise linear equations," Math. Oper. Res. 1, 1-27 (1976).

17. M.L. Fisher and F.J. Gould, "A simplicial algorithm for the nonlinear complementarity problem," Math. Program. 6, 281-300 (1974).

18. M.L. Fisher, F.J. Gould and J.W. Tolle, "A new simplicial approximation algorithm with restarts: relations between convergence and labelling," in Fixed points: algorithms and applications, S. Karamardian, ed., Academic Press, New York (1977).

19. W. Forster (ed.), Numerical solutions of highly nonlinear problems, North-Holland, Amsterdam (1980).

20. J. Freidenfelds, "Fixed-point algorithms and almost-complementary sets," TR71-17, Operations Research House, Stanford University, Stanford, California (1971).

21. R.M. Freund, "Variable-dimension complexes with applications," TR SOL 80-11, Department of Operations Research, Stanford University, Stanford, California (1980).

22. R.M. Freund and M.J. Todd, "A constructive proof of Tucker's combinatorial lemma," J. Comb. Theory (Series A) 30, 321-325 (1981).

23. C.B. Garcia, "A global existence theorem for the equation $Fx = y$," Center for Mathematical Studies in Business and Economics Report 7527, University of Chicago, Chicago, Illinois (1975).

24. ______ "A fixed point theorem including the last theorem of Poincare," Math. Program. 9, 227-239 (1975).

25. C.B. Garcia and F.J. Gould, "Relations between several path following algorithms and local and global Newton methods," SIAM Rev. 22, 263-274 (1980).

26. C.B. Garcia and W.I. Zangwill, "Determining all solutions to certain systems of nonlinear equations," Math. Oper. Res. 4, 1-14 (1979).

27. ______ "Finding all solutions to polynomial systems and other systems of equations," Math. Program. 16, 159-176 (1979).

28. W. Gochet, E. Loute and D. Solow, "Comparative computer results of three algorithms for solving prototype geometric programming problems," Cahiers du Centre d'Etudes de Recherche Operationelle 16, 469-486 (1974).

29. S.M. Goldfeld, R.E. Quandt and H.F. Trotter, "Maximization by quadratic hill-climbing," Econometrica 34, 541-551 (1966).

30. F.J. Gould and J.W. Tolle, "An existence theorem for solutions to $f(x) = 0$," Math. Program. 11, 252-262 (1976).

31. S. Karamardian (ed.), Fixed points: algorithms and applications, Proc. Conf. on Computing Fixed Points with Applications, Clemson University, Academic Press, New York (1977).

32. J. Katzenelson, "An algorithm for solving nonlinear resistive networks," Bell System Technical Journal 44, 1605-1620 (1965).

33. H.B. Keller, "Global homotopies and Newton methods," in Recent Advances in Numerical Analysis, C. de Boor and G.H. Golub, eds., Academic Press, New York (1978).

34. R.B. Kellogg, T.Y. Li and J. Yorke, "A constructive proof of the Brouwer fixed point theorem and computational results," SIAM J. Numer. Anal. 4, 473-483 (1976).

35. M. Kojima, "Computational methods for solving the nonlinear complementarity problem," Keio Engineering Reports 27, 1-41 (1974).

36. ______ "On the homotopic approach to systems of equations with separable mappings," Math. Program. Study 7, M.L. Balinski and R.W. Cottle, eds., North Holland, Amsterdam, 170-184 (1978).

37. ______ "An introduction to variable dimension algorithms for solving systems of equations," in Numerical Solution of Nonlinear Equations, E.L. Allgower, K. Glashoff and H.-O. Peitgen, eds., Springer-Verlag, Berlin (1981).

38. M. Kojima, H. Nishino and N. Arima, "A PL homotopy for finding all the roots of a polynomial," Math. Program. 16, 37-62 (1979).

39. M. Kojima and Y. Yamamoto, "Variable dimension algorithms, part I: basic theory," Research Report B-77, Department of Information Sciences, Tokyo Institute of Technology, Tokyo (1979).

40. ______ "Variable dimension algorithms, part II: some new algorithms and triangulations with continuous refinement of mesh size," Research Report B-82, Department of Information Sciences, Tokyo Institute of Technology, Tokyo (1980).

41. H.W. Kuhn, "Simplicial approximation of fixed points," Proc. Nat. Acad. Sci., U.S.A., 61, 1238-1242 (1968).

42. ______ "Approximate search for fixed points," Computing Methods in Optimization Problems 2, Academic Press, New York (1969).

43. ______ "Finding roots by pivoting," Fixed Points: Algorithms and Applications, S. Karamardian, ed., Academic Press, New York, 11-39 (1977).

44. H. Kuhn and J.G. Mackinnon, "Sandwich method for finding fixed points," J. Optimization Theory Appl. 17, 189-204 (1975).

45. G. van der Laan and A.J.J. Talman, "A restart algorithm for computing fixed points without an extra dimension," Math. Program. 17, 74-84 (1979).

46. ______ "A restart algorithm without an artificial level for computing fixed points on unbounded regions," in Functional Differential Equations and Approximation of Fixed Points, H.-O. Peitgen and H.-O. Walther, eds., Lect. Notes Math. 730, Springer-Verlag, Berlin (1979).

47. ______ "Interpretation of the variable dimension fixed point algorithm with an artificial level," Faculty of Econometrics, Free University, Amsterdam (1979).

48. ______ "An improvement of fixed point algorithms by using a good triangulation," Math. Program. 18, 274-285 (1980).

49. ______ "A new subdivision for computing fixed points with a homotopy algorithm," Math. Program. 19, 78-91 (1980).

50. ______ "Convergence and properties of recent variable dimension algorithms," in Numerical Solution of Highly Nonlinear Problems, W. Forster, ed., North-Holland, Amsterdam (1980).

51. ______ "A class of simplicial restart fixed point algorithms without an extra dimension," Math. Program. 20, 33-48 (1981).

52. C.E. Lemke, "A survey of complementarity theory," in Variational Inequalities and Complementarity Problems, R.W. Cottle, F. Gianessi and J.-L. Lions, eds., John Wiley, Chichester (1980).

53. C.E. Lemke and J.T. Howson, Jr., "Equilibrium points of bimatrix games," SIAM J. Appl. Math. 12, 413-423 (1964).

54. H.J. Luthi, Komplementaritats-und Fixpunktalgorithmen in der mathematischen Programmierung, Spieltheorie und Okonomie, Lect. Notes Econ. Math. Syst. 129, Springer-Verlag, Berlin (1976).

55. O.H. Merrill, "Applications and extensions of an algorithm that computes fixed points of a certain upper semi-continuous point to set mappings," Ph.D. Thesis, Dept. of Industrial Engineering, University of Michigan (1972).

56. M.D. Meyerson and A.H. Wright, "A new and constructive proof of the Borsuk-Ulam theorem," Proc. Amer. Math. Soc. 73, 134-136 (1979).

57. A.N. Netravali and R. Saigal, "Optimum quantizer design using a fixed-point algorithm," The Bell System Technical Journal 55, 1423-1435 (1976).

58. J.M. Ortega and W.C. Rheinboldt, Iterative solutions of nonlinear equations in several variables, Academic Press, New York-London (1970).

59. H.-O. Peitgen and H.-O. Walther, eds., Functional differential equations and approximation of fixed points, Lect. Notes Math. 730, Springer-Verlag, Berlin (1979).

60. P.M. Reiser, "A modified integer labeling for complementarity algorithms," Math. Oper. Res. 6, 129-139 (1981).

61. S.M. Robinson, ed., Analysis and computation of fixed points, Academic Press, New York (1980).

62. R. Saigal, "On paths generated by fixed point algorithms," Math. Oper. Res. 1, 359-380 (1976).

63. ______ "Fixed point computing methods," Encyclopedia of Computer Science and Technology, 8, Marcel Dekker Inc., New York (1977).

64. ______ "On the convergence rate of algorithms for solving equations that are based on complementary pivoting," Math. Oper. Res. 2, 108-124 (1977).

65. ______ "The fixed point approach to nonlinear programming," Math. Program. Stud. 10, 142-157 (1979).

66. ______ "Efficient algorithms for computing fixed points when mappings may be separable," manuscript, Department of Industrial Engineering and Management Sciences, Northwestern University, Evanston, Illinois (1979).

67. R. Saigal and M.J. Todd, "Efficient acceleration techniques for fixed point algorithms," SIAM J. Numer. Anal. 15, 997-1007 (1978).

68. H. Scarf, "The core of an N-person game," Econometrica 35, 50-69 (1967).

69. ______ "The approximation of fixed points of a continuous mapping," SIAM J. Appl. Math. 15, 1328-1343 (1967).

70. ______ "The computation of equilibrium prices," to appear in the proceedings of a conference on Applied General Equilibrium Analysis held in San Diego, California (1981).

71. H.E. Scarf with T. Hansen, Computation of economic equilibria, Yale University Press, New Haven, Connecticut (1973).

72. S. Shamir, "Two new triangulations for homotopy fixed point algorithms with an arbitrary grid refinement," in Analysis and Computation of Fixed Points, S.M. Robinson, ed., Academic Press, New York (1980).

73. L.S. Shapley, "On balanced games without side payments," in Mathematical Programming, T.C. Hu and S.M. Robinson, eds., Academic Press, New York (1972).

74. S. Smale, "A convergent process of price adjustment and global Newton methods," J. Math. Econ. 3, 1-14 (1976).

75. D. Solow, "Comparative computer results of a new complementary pivot algorithm for solving equality and inequality constrained optimization problems," Math. Program. 18, 169-185 (1980).

76. M.J. Todd, The computation of fixed points and applications, Lect. Notes Econ. Math. Syst. 124, Springer-Verlag, Berlin (1976).

77. ______ "On triangulations for computing fixed points," Math. Program. 10, 322-346 (1976).

78. ______ "Orientation in complementary pivoting," Math. Oper. Res. 1, 54-66 (1976).

79. ______ "Union Jack triangulations," Fixed Points: Algorithms and Applications, S. Karamardian, ed., Academic Press, New York (1977).

80. ______ "Improving the convergence of fixed-point algorithms, Math. Program. Stud. 7, M.L. Balinski and R.W. Cottle, eds., North Holland, Amsterdam, 151-169 (1978).

81. ______ "Fixed-point algorithms that allow restarting without an extra dimension," School of Operations Research and Industrial Engineering, College of Engineering, Cornell University, Ithaca, NY, Technical Report No. 379 (1978).

82. ______ "Optimal dissection of simplices," SIAM J. Appl. Math. 34, 792-803 (1978).

83. ______ "Global and local convergence and monotonicity results for a recent variable-dimension simplicial algorithm," in Numerical Solution of Highly Nonlinear Problems, W. Forster, ed., North-Holland, Amsterdam (1980).

84. ______ "Exploiting structure in piecewise-linear homotopy algorithms for solving equations," Math. Program. 18, 233-247 (1980).

85. ______ "Traversing large pieces of linearity in algorithms that solve equations by following piecewise-linear paths," Math. Oper. Res. 5, 242-257 (1980).

86. ______ "Numerical stability and sparsity in piecewise-linear algorithms," in Analysis and Computation of Fixed Points, S.M. Robinson, ed., Academic Press, New York (1980).

87. ______ "On the computational complexity of piecewise-linear homotopy algorithms," manuscript, Department of Applied Mathematics and Theoretical Physics, University of Cambridge, Cambridge (1981).

88. ______ "Efficient methods of computing economic equilibria," to appear in the proceedings of a conference on Applied General Equilibrium Analysis held in San Diego, California (1981).

89. ______ "Piecewise-linear homotopy algorithms for sparse systems of nonlinear equations," Technical Report, School of Operations Research and Industrial Engineering, Cornell University, Ithaca, New York (1981).

90. ______ "PLALGO: a FORTRAN implementation of a piecewise-linear homotopy algorithm for solving systems of nonlinear equations," Technical Report (revised), School of Operations Research and Industrial Engineering, Cornell University, Ithaca, New York (1981).

91. M.J. Todd and R. Acar, "A note on optimally dissecting simplices," Math. Oper. Res. 5, 63-66 (1980).

92. M.J. Todd and A.H. Wright, "A variable-dimension simplicial algorithm for antipodal fixed-point theorems," Numer. Funct. Anal. Optimization 2, 155-186 (1980).

93. H. Tuy, Ng. v. Thoai and L.d. Muu, "A modification of Scarf's algorithm allowing restarting," Math. Operationsforsch. Statist., Ser. Optimization 9, 357-372 (1978).

94. L. van der Heyden, "Restricted primitive sets in a regularly distributed list of vectors and simplicial subdivisions with arbitrary refinement factors," Discussion Paper Series No. 79D, Kennedy School of Government, Harvard University, Cambridge, Massachusetts (1980).

95. H. Wacker, ed., Continuation methods, Academic Press, New York (1978).

96. L.T. Watson, "Computational experience with the Chow-Yorke algorithm," Math. Program. 19, 92-101 (1980).

97. L.T. Watson and D. Fenner, "Chow-Yorke algorithm for fixed points or zeros of C^2 maps," ACM Trans. Math. Software 6, 252-260 (1980).

98. A.H. Wright, "The octahedral algorithm, a new simplicial fixed point algorithm," Math. Program. 21, 47-69 (1981).

Vol. 873: Constructive Mathematics, Proceedings, 1980. Edited by F. Richman. VII, 347 pages. 1981.

Vol. 874: Abelian Group Theory. Proceedings, 1981. Edited by R. Göbel and E. Walker. XXI, 447 pages. 1981.

Vol. 875: H. Zieschang, Finite Groups of Mapping Classes of Surfaces. VIII, 340 pages. 1981.

Vol. 876: J. P. Bickel, N. El Karoui and M. Yor. Ecole d'Eté de Probabilités de Saint-Flour IX – 1979. Edited by P. L. Hennequin. XI, 280 pages. 1981.

Vol. 877: J. Erven, B.-J. Falkowski, Low Order Cohomology and Applications. VI, 126 pages. 1981.

Vol. 878: Numerical Solution of Nonlinear Equations. Proceedings, 1980. Edited by E. L. Allgower, K. Glashoff, and H.-O. Peitgen. XIV, 440 pages. 1981.

Vol. 879: V. V. Sazonov, Normal Approximation – Some Recent Advances. VII, 105 pages. 1981.

Vol. 880: Non Commutative Harmonic Analysis and Lie Groups. Proceedings, 1980. Edited by J. Carmona and M. Vergne. IV, 553 pages. 1981.

Vol. 881: R. Lutz, M. Goze, Nonstandard Analysis. XIV, 261 pages. 1981.

Vol. 882: Integral Representations and Applications. Proceedings, 1980. Edited by K. Roggenkamp. XII, 479 pages. 1981.

Vol. 883: Cylindric Set Algebras. By L. Henkin, J. D. Monk, A. Tarski, H. Andréka, and I. Németi. VII, 323 pages. 1981.

Vol. 884: Combinatorial Mathematics VIII. Proceedings, 1980. Edited by K. L. McAvaney. XIII, 359 pages. 1981.

Vol. 885: Combinatorics and Graph Theory. Edited by S. B. Rao. Proceedings, 1980. VII, 500 pages. 1981.

Vol. 886: Fixed Point Theory. Proceedings, 1980. Edited by E. Fadell and G. Fournier. XII, 511 pages. 1981.

Vol. 887: F. van Oystaeyen, A. Verschoren, Non-commutative Algebraic Geometry, VI, 404 pages. 1981.

Vol. 888: Padé Approximation and its Applications. Proceedings, 1980. Edited by M. G. de Bruin and H. van Rossum. VI, 383 pages. 1981.

Vol. 889: J. Bourgain, New Classes of $\mathcal{L}^p$-Spaces. V, 143 pages. 1981.

Vol. 890: Model Theory and Arithmetic. Proceedings, 1979/80. Edited by C. Berline, K. McAloon, and J.-P. Ressayre. VI, 306 pages. 1981.

Vol. 891: Logic Symposia, Hakone, 1979, 1980. Proceedings, 1979, 1980. Edited by G. H. Müller, G. Takeuti, and T. Tugué. XI, 394 pages. 1981.

Vol. 892: H. Cajar, Billingsley Dimension in Probability Spaces. III, 106 pages. 1981.

Vol. 893: Geometries and Groups. Proceedings. Edited by M. Aigner and D. Jungnickel. X, 250 pages. 1981.

Vol. 894: Geometry Symposium. Utrecht 1980, Proceedings. Edited by E. Looijenga, D. Siersma, and F. Takens. V, 153 pages. 1981.

Vol. 895: J.A. Hillman, Alexander Ideals of Links. V, 178 pages. 1981.

Vol. 896: B. Angéniol, Familles de Cycles Algébriques – Schéma de Chow. VI, 140 pages. 1981.

Vol. 897: W. Buchholz, S. Feferman, W. Pohlers, W. Sieg, Iterated Inductive Definitions and Subsystems of Analysis: Recent Proof-Theoretical Studies. V, 383 pages. 1981.

Vol. 898: Dynamical Systems and Turbulence, Warwick, 1980. Proceedings. Edited by D. Rand and L.-S. Young. VI, 390 pages. 1981.

Vol. 899: Analytic Number Theory. Proceedings, 1980. Edited by M.I. Knopp. X, 478 pages. 1981.

Vol. 900: P. Deligne, J. S. Milne, A. Ogus, and K.-Y. Shih, Hodge Cycles, Motives, and Shimura Varieties. V, 414 pages. 1982.

Vol. 901: Séminaire Bourbaki vol. 1980/81 Exposés 561–578. III, 299 pages. 1981.

Vol. 902: F. Dumortier, P.R. Rodrigues, and R. Roussarie, Germs of Diffeomorphisms in the Plane. IV, 197 pages. 1981.

Vol. 903: Representations of Algebras. Proceedings, 1980. Edited by M. Auslander and E. Lluis. XV, 371 pages. 1981.

Vol. 904: K. Donner, Extension of Positive Operators and Korovkin Theorems. XII, 182 pages. 1982.

Vol. 905: Differential Geometric Methods in Mathematical Physics. Proceedings, 1980. Edited by H.-D. Doebner, S.J. Andersson, and H.R. Petry. VI, 309 pages. 1982.

Vol. 906: Séminaire de Théorie du Potentiel, Paris, No. 6. Proceedings. Edité par F. Hirsch et G. Mokobodzki. IV, 328 pages. 1982.

Vol. 907: P. Schenzel, Dualisierende Komplexe in der lokalen Algebra und Buchsbaum-Ringe. VII, 161 Seiten. 1982.

Vol. 908: Harmonic Analysis. Proceedings, 1981. Edited by F. Ricci and G. Weiss. V, 325 pages. 1982.

Vol. 909: Numerical Analysis. Proceedings, 1981. Edited by J.P. Hennart. VII, 247 pages. 1982.

Vol. 910: S.S. Abhyankar, Weighted Expansions for Canonical Desingularization. VII, 236 pages. 1982.

Vol. 911: O.G. Jørsboe, L. Mejlbro, The Carleson-Hunt Theorem on Fourier Series. IV, 123 pages. 1982.

Vol. 912: Numerical Analysis. Proceedings, 1981. Edited by G. A. Watson. XIII, 245 pages. 1982.

Vol. 913: O. Tammi, Extremum Problems for Bounded Univalent Functions II. VI, 168 pages. 1982.

Vol. 914: M. L. Warshauer, The Witt Group of Degree k Maps and Asymmetric Inner Product Spaces. IV, 269 pages. 1982.

Vol. 915: Categorical Aspects of Topology and Analysis. Proceedings, 1981. Edited by B. Banaschewski. XI, 385 pages. 1982.

Vol. 916: K.-U. Grusa, Zweidimensionale, interpolierende Lg-Splines und ihre Anwendungen. VIII, 238 Seiten. 1982.

Vol. 917: Brauer Groups in Ring Theory and Algebraic Geometry. Proceedings, 1981. Edited by F. van Oystaeyen and A. Verschoren. VIII, 300 pages. 1982.

Vol. 918: Z. Semadeni, Schauder Bases in Banach Spaces of Continuous Functions. V, 136 pages. 1982.

Vol. 919: Séminaire Pierre Lelong – Henri Skoda (Analyse) Années 1980/81 et Colloque de Wimereux, Mai 1981. Proceedings. Edité par P. Lelong et H. Skoda. VII, 383 pages. 1982.

Vol. 920: Séminaire de Probabilités XVI, 1980/81. Proceedings Edité par J. Azéma et M. Yor. V, 622 pages. 1982.

Vol. 921: Séminaire de Probabilités XVI, 1980/81. Supplément: Géométrie Différentielle Stochastique. Proceedings. Edité par J. Azéma et M. Yor. III, 285 pages. 1982.

Vol. 922: B. Dacorogna, Weak Continuity and Weak Lower Semicontinuity of Non-Linear Functionals. V, 120 pages. 1982.

Vol. 923: Functional Analysis in Markov Processes. Proceedings 1981. Edited by M. Fukushima. V, 307 pages. 1982.

Vol. 924: Séminaire d'Algèbre Paul Dubreil et Marie-Paule Malliavin Proceedings, 1981. Edité par M.-P. Malliavin. V, 461 pages. 1982

Vol. 925: The Riemann Problem, Complete Integrability and Arithmetic Applications. Proceedings, 1979-1980. Edited by D. Chudnovsky and G. Chudnovsky. VI, 373 pages. 1982.

Vol. 926: Geometric Techniques in Gauge Theories. Proceedings 1981. Edited by R. Martini and E.M.de Jager. IX, 219 pages. 1982